Theodore

Augsburg College
George Sverdrup Library
Minneapolis, Minnesota 55404

THE VINLAND VOYAGES

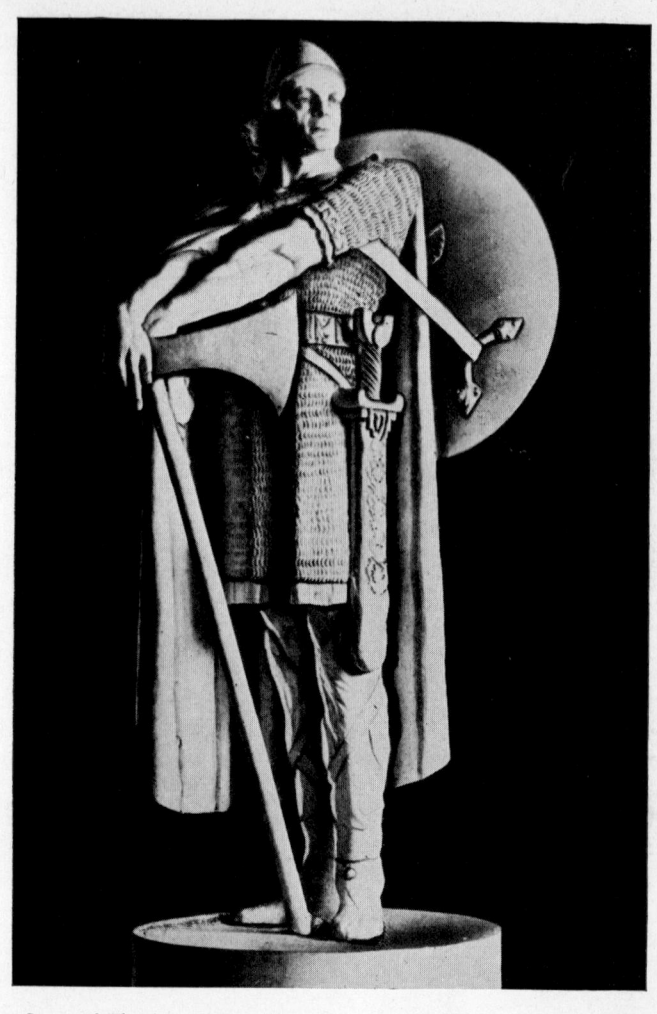

Statue of Thorfinn Karlsefni by Einar Jónsson in Fairmount Park, Philadelphia. (From *Einar Jónsson Myndir*, Copenhagen, 1925.)

AMERICAN GEOGRAPHICAL SOCIETY
RESEARCH SERIES NO. 18
Edited by HALLDÓR HERMANNSSON

THE VINLAND VOYAGES

BY

MATTHIAS THÓRDARSON
Director, National Museum of Iceland

TRANSLATED BY
THORSTINA JACKSON WALTERS

WITH AN INTRODUCTION BY
VILHJÁLMUR STEFÁNSSON

AMERICAN GEOGRAPHICAL SOCIETY
BROADWAY AT 156TH STREET
NEW YORK
1930

COPYRIGHT, 1930
BY
THE AMERICAN GEOGRAPHICAL SOCIETY
OF NEW YORK

George Grady Press
New York

CONTENTS

	PAGE
INTRODUCTION, BY VILHJÁLMUR STEFÁNSSON	v
THE DOCUMENTARY SOURCES	1
LEIF ERICSSON'S VOYAGE IN THE YEAR 1000	5
THE UNSUCCESSFUL VOYAGE OF THORSTEIN, LEIF'S BROTHER, IN 1001	9
THORFINN KARLSEFNI'S COLONIZATION EXPEDITION, 1003-1006(?)	11
THE VOYAGE FROM GREENLAND TO STREAMFJORD, 1003	12
THORHALL THE HUNTER'S SEPARATE QUEST NORTHWARD, 1004	31
THORFINN'S VOYAGE SOUTHWARD TO HÓP, 1004	33
THE RETURN TO STREAMFJORD AND SEARCH FOR THORHALL, 1005(?)	56
THE RETURN TO GREENLAND, 1006(?)	61
CONCLUSION	65
INDEX	69

LIST OF ILLUSTRATIONS

FIG.		PAGE
	Statue of Thorfinn Karlsefni	Frontispiece
1	The Horn, Iceland, near where Eric the Red had his home	facing 2
2	Nidaros (Trondhjem), whence Leif Ericsson set out	facing 2
3	Norse ruins in Greenland	facing 6
4	Ericsfjord in the Eastern Settlement of Greenland	facing 6
5	Map showing conjectural route of Leif Ericsson's voyage in the year 1000	8
6	Facsimile from Hauk's Book of the story of Leif's finding of Vinland	facing 10
7	Lysufjord in the Western Settlement of Greenland	facing 10
8	Map showing conjectural route of Thorfinn Karlsefni's voyages, 1003-1006(?)	17
9	Coast of Labrador near Cape Mugford	facing 18
10	Coast of Labrador near Cape Mugford	facing 18
11	Map showing in greater detail Thorfinn Karlsefni's conjectural route in the Gulf of St. Lawrence	20
12	Western coast of Newfoundland	facing 22
13	Northeastern coast of Newfoundland	facing 22
14	Cape Gaspé: the extreme point of the cape	facing 24
15	Cape Gaspé: a secondary headland	facing 24
16	Head of Chaleur Bay	facing 28
17	Bird ledges, Gaspé Peninsula	facing 28
18	Percé Rock, Gaspé Peninsula	facing 30
19	Facsimile from Hauk's Book of the description of Hóp	facing 34
20	Bluffs at Highland Light, Cape Cod	facing 44
21	Wild rice (*Zizania aquatica*)	facing 44
22	Viking battle-ax dating from the year 1000	facing 66
23	Viking swords of the same period	facing 66

INTRODUCTION

By Vilhjálmur Stefánsson

A logical introduction to the discovery of the American continent by Icelanders or Greenlanders, as related in this book, is a brief account of the steps by which the discoverers reached the vantage points of Iceland and Greenland.

The striking upheaval of folk energy known as the Viking Age is the background of our story. The books about that period increase, but we grow thereby more in wealth of theory than of fact. It is only a slight overstatement to say that the more we know about the doings of the time the less confident we are about the theories that explain the way things were done.

But it is clear that during half a dozen centuries preceding 1000 A.D. the caldron of Norse energy boiled over again and again. We have conceived that age and that part of the world as swarming with restless and brawling freebooters who made aimless if joyous raids to south and west, the waves of them ebbing in France or the British Isles, the energy dying down slowly as they became civilized and Christian. The picture as retained by learned Europe has been so uniformly unfavorable that few until recently have suspected possible inaccuracies.

True, it has been said with increasing frequency that the Norsemen may not have been quite so black as they were painted, since practically every stroke in their portraits had been laid on by the clergy, who, as the only learned men of Christian Europe, were the historians of the Viking Age and whose animosity was particularly violent because the monasteries and churches had been the main targets of depredation.

In the last fifty years it seems to have been growing clear at last that the civilization of the Vikings was high, though different from the Christian. As a brawling horde of ruffians they could, for instance, be thought of as overrunning and plundering northern France; but without a well-developed machinery of organization they could not have welded their conquests into that disciplined and powerful Normandy which presently conquered England. It is gratifying to French vanity if you can believe, though it is not easy to believe, that the Norsemen brought with them to the continent no marked intellectual or cultural gifts except an ability to learn from the conquered how they should behave as conquerors.

It is incredible that men with such quick ability to learn should have learned nothing of the ways of a high civilization till they met the Christians.

INTRODUCTION

But even with an indigenous or at least non-Christian civilization postulated for the Norsemen, their migrations and doings are still difficult to explain. Having insisted on the difficulty, we give it up as a bad job and pass to a rapid sketch of the things that appear to have a significant bearing on American discovery.

With regard to that most striking, or at least most advertised, of Viking abilities, the navigation of the high seas, we may profit by the great antiquities that are becoming popular in science and scholarship. During the last twenty years the twenty million years of astronomy have grown to beyond two hundred million. In sixty years the six thousand years of human history have been stretched well beyond six hundred thousand. Then why not add a good thousand years to that schoolroom estimate of twenty years ago that pictured the Phoenicians as pioneers in seamanship when they were dodging cautiously north from bay to haven along the coasts of Europe till they found tin in Britain?

According to that textbook view the sailing of the high seas was unknown until the Vikings began swarming the ocean as well as the lands, somewhere around 700 A.D. Applying the general principle that a short estimate of time is inherently improbable, we find a key to Viking civilization in what may be inferred from that fragmentary account of Pytheas' voyages that has been saved to us, where, as Nansen has maintained, it seems probable that when he arrived at the north tip of Scotland, somewhere around 325 B.C., the people there told Pytheas that if he were to continue a certain number of days in the same direction (north or northeast) he would come to another country which they were able to describe to him. This would mean that in the days when Alexander was conquering Asia the Scots were sailing the high seas; for, under normal conditions of wind and weather, it would happen again and again on voyages between Scotland and Norway that ships would be out of sight of land for several days.

Or was it the Norwegians who were already sailing to Scotland a thousand years before the nominal beginning of the Viking Age? More likely it was both, for when you have across a navigable water two peoples of the known traits of the Scots and the Norwegians you would have to assume that one would quickly learn from the other.

This makes (if possible) more absurd the inherently absurd assumption that in the early Viking Age the Irish had no boats except coracles. For if the Scots sailed to Norway, or if the Norwegians sailed to Scotland, either or both would sail to Ireland, and the knowledge of shipping would spread. The Irish would

INTRODUCTION

acquire real ships, or at least boats, and would sail or row in them to Iceland, not in coracles.

We know from two independent sources that the Irish did discover Iceland before 800 A.D. The first source is Irish, the work of the monk Dicuil who tells, writing somewhere around 825 A.D., that the Irish had been in Iceland thirty years before. Or he may mean they had been there longer, for there is an ambiguity in his statement which may be interpreted to say either that the Irish had been in the Faeroes a hundred years and in Iceland thirty, or else that they had been in both islands a hundred years and that the mentioned visit of thirty years before to Iceland was merely a particular voyage in which Dicuil was interested.

Dicuil's work has internal evidence of its truth. He is writing in part to contradict a popular erroneous belief about the length and character of the summer day in the Far North. In Iceland (on the south coast), says he, the truth is that you never see the sun through its full circle, but you have, nevertheless, so bright a twilight in midsummer that even at midnight you could take off your shirt and, without a candle, pick the lice from it. Today we should convey the same idea by saying that on the southern coast of Iceland in midsummer you can read a newspaper at midnight, though the sun itself is hidden.

It is not conceivable that a man in Ireland should know so precisely this distinction between direct sunlight and bright twilight on the south coast of Iceland without either having been there himself or, as Dicuil expressly states, possessing the testimony of men who had been there.

About the time when the Irish began to visit Iceland in a type of ship that must have been more efficient if not more seaworthy than coracles, there were a dozen or a score of independent kings and earls in Norway. Then, after the middle of the ninth century, one of the petty kings named Harald the Fairhaired developed a military and organizing genius, overthrew his fellow kings, and welded the country into a single monarchy. Princelings who had lived by taxation or its equivalent did not find it pleasant to take orders and pay tribute. This brought about what has been described as the only large-scale migration in history where the nobility moved out and the peasantry remained at home. But it is to be remembered that the emigrating Norse nobility were accompanied by their retainers and also by their slaves, so that, while a great many of the upper classes were involved, there were also concerned numbers of every grade below them.

The biggest lot of expatriates were those who conquered France to become the rulers of Normandy and the lords of England. Other large contingents went to the British Isles direct, conquering, among other places, Yorkshire and Lancashire in England, Sutherland and Caithness in Scotland, and the Dublin region of Ireland.

If any people in Europe have been more arrogant than the Romans of two thousand years ago or the English of the last two hundred years, it was the Vikings of a thousand years ago. From their point of view nothing was discovered till they discovered it, and so they were not fabricating but rather speaking in tune with their time when they said that they discovered Iceland around 850 A.D. But for long periods before they had been in intimate if predatory contact with Ireland, and they were masters there now. They were apt at learning, if perhaps not quite so apt as the French have asserted. Nothing was more familiar to them than the lore of the sea, and it is simply not credible that they could have remained in such long and intimate contact with the Irish without learning from them that there were other Irishmen resident in Iceland with either intermittent or regular navigation between the two countries. Even if it be true that the Norse Naddodd "discovered" Iceland accidentally when driven out of his Faeroes course by a storm (as the sagas assert), the news of the "discovery" can have been little more of a surprise to the Vikings in Ireland than the Columbian "discovery" of America was to the learned men of the Papacy six centuries later.

Just as clearly as Dicuil states the presence of the Irish in Iceland before 825 A.D. do the Norsemen themselves state that they found the Irish in Iceland when they arrived.

In modern times the Irish have built up a reputation as a turbulent and even a warlike people, but they were comparatively gentle and peaceable in the Viking Age. They had, however, offered partially effective resistance to the Norse conquest in Ireland; but they were in a position to offer none in Iceland. This was because they were so few, but more because those few were, in the main, clergy with their retainers who had gone to Iceland for solitude, and not warriors. By the Norse accounts many of the Christian Irish fled the country around or after 870 A.D.

Between 870 and 930 A.D. Iceland is believed to have acquired a population of about 50,000. The language, institutions, and the dominant nobility were all Norse; but the people did not all come to Iceland direct from the Norse countries. Many came instead by way of the British Isles, where they had visited for years or a decade and where some families had stayed for two or three generations. Most of the

INTRODUCTION

chieftains had Norse wives, but some had Irish. Besides, there was occasional polygamy with Irish second wives, and there was concubinage. Many settlers who came to Iceland were voluntarily accompanied by whole families of Irish. There was slavery, and many of the slaves were Irish.

All this brought in a considerable percentage of Irish blood. During the next few centuries there was a good deal of commerce with Europe, a considerable part of it with Ireland, and the traders frequently settled in Iceland. The percentage of various bloods is accordingly a much disputed question. It has been argued that the proportion may have been something like 60 per cent Norwegian, 30 per cent Irish, and the remaining 10 per cent made up from the various North European nationalities, perhaps in the following order: Scots, English, Danes, Swedes. Most scholars feel, however, that the Scottish-Irish percentage was less than 30, perhaps 20 or even as low as 10.

By 925 A.D. the need for a central government was keenly felt in Iceland. A chieftain named Ulfljot went to Norway to study political forms and procedure, returned to Iceland, and made his report; and the first parliament of the republic of Iceland met in session at what was later called Thingvellir in 930 A.D. The New World, destined to become so peculiarly a home of republics, was through this action destined to a discovery by the citizens of a republic.

The eventual discovery of the North American mainland hinges upon a fashionable practice of the day, that of man killing, which, like cocktail shaking in the later America, was against the law but was indulged in by the best people. A Norwegian, Thorvald Asvaldsson, killed too many of his neighbors, or at least killed some who had too much influence, and was outlawed. He went abroad with his family and settled in Iceland.

In this exiled family was a boy with red hair named Eric. When he grew up in Iceland, where man killings were quite as fashionable as in Norway, he, like his father, killed people that were too influential and in 982 was exiled for a three-year period.

There was a rumor that about eighty years before this a sailor named Gunnbjörn had seen some skerries in the ocean to the west of Iceland. Instead of commonplace outlawry spent perhaps in the British Isles, Eric the Red decided to sail in quest of these skerries, discovered or rediscovered Greenland, and explored its west coast for the three years of his sentence.

During the third year of exploration, as the saga tells us, Eric formed a plan of colonization. He had a genius for advertising that made him prophetically American. For the narrative says he conceived that

INTRODUCTION

people would all the more desire to colonize the new country if it had an attractive name, and so he called it the Green Land.

When Eric returned to Iceland, he "sold" the new country so attractively to his neighbors that he was accompanied back to Greenland, in 985 or 986, by fourteen ships carrying between four and five hundred persons and with them horses, cattle, sheep, goats, chickens, dogs, cats, and varied household goods.

The colony had probably grown to a population of a thousand, with the nucleus of a parliamentary government resembling that of Iceland, by the year 999, at which point the author of the book we are introducing takes up the story of Greenland in connection with that of the discovery of the North American mainland.

THE VINLAND VOYAGES[1]

THE year 1000 is a veritable milestone in the history of the Norse nations, associated with events of deep-lying results whose significance has been felt through all the subsequent centuries.

It was early in the summer of that year that King Olaf Tryggvason of Norway bade a last farewell to his three guests of many months, the young Icelandic chieftains Gissur Teitsson, Hjalti Skeggjason, and Leif Ericsson. The last-named hailed from the Icelandic colony in Greenland and was the son of the brave chief Eric the Red. Leif was born and brought up in Iceland but had spent a number of years with his father in Greenland. These three young men had been converted to Christianity at King Olaf's court and had made a solemn vow to him to act as missionaries in their respective homelands, Gissur and Hjalti in Iceland and Leif in Greenland.

The three young chieftains set out to sea, but King Olaf left a short time after for the Land of the Wends, on the southern shores of the Baltic, and was

[1] The original paper of which this is a translation was published in substantially the same form in *Safn Til Sögu Islands og Islenzkra Bókmenta*, Vol. 6, No. 1, Reykjavík, 1929.

destined never to see Norway again. He met his fate at the battle of the Island of Svold, an event of marked importance in Norse history. There the kings of Sweden and Denmark attacked him, and Norway's warrior-missionary lost his life. His Icelandic friends made history that summer also, Hjalti and Gissur succeeding in persuading the Althing of Iceland to adopt Christianity as the religion of the country, while Leif Ericsson had, before the end of the summer, discovered a New World.

In the following pages an analysis is attempted of Leif Ericsson's discovery and related events and adventures. The story of his discovery and that of his countrymen in America is especially known from a collection of various sagas made some six centuries ago by Hauk Erlendsson, an Icelandic lawman. This work is known as "Hauksbók" (Hauk's Book); one of the important portions of it is the "Saga of Eric the Red." It derives its name from Eric, Leif's father, and there one finds the clearest and most reliable accounts of Vinland. The original manuscript of Hauk's Book is in the Arna-Magnaean Collection in Copenhagen, Number 544, 4to; and there is another vellum manuscript of the Eric Saga in the same collection, Number 557, 4to. The latter is written at least a century later than the former, and the scribe's name is unknown. Arni Magnússon acquired the second manuscript from Bishop

Fig. 1—The Horn, northernmost part of Iceland, near which Eric the Red had his home and where Leif Ericsson probably was born.

Fig. 2—A headland at the entrance to the harbor of Nidaros (Trondhjem), Norway, the town from which Leif set out on his voyage of discovery. (Courtesy of Norwegian Government Railways.)

DOCUMENTARY SOURCES

Jón Vídalín, and it had probably been the property of the Skálholt Cathedral in Iceland. The two texts vary somewhat, and by comparison it becomes evident that Hauk has made some additions to the original text, while the writer of the other text has made many changes in the phraseology. On the whole, the narrative is more original and better arranged in Hauk's version than in the other manuscript; that, at least, is the opinion of Professor Finnur Jónsson.[2]

It is probable that the Saga of Eric the Red was originally written long before Hauk's time. At the end of the Saga there is a reference to Brand Sæmundsson, Bishop of Hólar; and he is there called Brand the First, which seems to imply, unless this is a later interpolation, that the Saga was written after the episcopate of Bishop Brand Jónsson, the second of that name, which lasted only two years, 1263-64. Professor Finnur Jónsson believes that the original story was written about 1200, or at least not later than the first quarter of the thirteenth century; but he concedes that some additions and changes were made during the last quarter of that century. Professor Gustav

[2] See his introduction to the *Hauksbók* (edited by Finnur Jónsson and Eiríkur Jónsson, Copenhagen, 1892-1896), pp. lxxxiii-lxxxiv, and his *Den oldnorske og oldislandske litteraturs historie*, Vol. 2, pp. 646-648, 1901. The two manuscripts of the Saga are reproduced in facsimile with printed text and English translation by A. M. Reeves, *The Finding of Wineland the Good*, London, 1890.

Storm, who has made the best edition of the Saga[3] and also written an important treatise on the subject,[4] expressed the opinion that the Saga was written in Snæfellsness, Iceland. It seems certain that the author was familiar with that part of the country, and there is every reason to believe that some portions of the Saga are traceable to Ari Thorgilsson the Learned, and probably also to Sturla Thórdarson, the historian. There may have been saga writers also in the Monastery at Helgafell during the first quarter of the thirteenth century such as the two abbots Ketil Hermundsson and Hall Gissursson. One may infer that much of the subject matter in the Saga was derived directly from the narratives of the two principal characters in it, Thorfinn Karlsefni and his wife, Gudrid. At the conclusion of the "Tale of the Greenlanders" there is a statement to the effect that Karlsefni had supplied most of the material relating to the Vinland voyages. This is probably true, although the veracity of the narrative, as set forth in the Tale, is doubtful, a matter that will be referred to later.

It must be noted that much material similar to that included in the Saga of Eric the Red is woven into

[3] *Eiríks saga rauda og Flatøbogens Grænlendingatháttur*, Copenhagen, 1891.

[4] In *Aarbøger for Nordisk Oldkyndighed og Historie*, 1887, pp. 293-372. An English version in *Mémoires de la Société Royale des Antiquaires du Nord*, 1884-89, pp. 307-370.

DOCUMENTARY SOURCES

the Saga of Olaf Tryggvason, particularly that written by Jón Thórdarson, the priest, in the renowned "Flatey Book."[5] It forms two interpolated portions, and they are called the "Tale of Eric the Red" and the "Tale of the Greenlanders" respectively.[6] It seems probable that they are based to some extent on the original "Saga of Eric the Red" either in its written or oral form. It is evident, however, that the tradition has undergone many changes, been added to and thus become less reliable; hence these tales are not to be depended upon whenever they contradict the Saga. Yet they cannot be disregarded entirely. The best edition of them is by Gustav Storm (together with the "Saga of Eric the Red") and both he and Finnur Jónsson agree as to their slight historical value.[7]

The narrative of Leif's voyage from Norway and his discoveries in the summer of 1000 as presented in the "Saga of Eric the Red," Hauk's Book text, is as follows:

[5] See *Flateyjarbók* (edited by G. Vigfusson and C. R. Unger, 3 vols., Christiania, 1860), Vol. 1, pp. 429-432 and 538-549. Those portions are given in facsimile with English translation in A. M. Reeves, *The Finding of Wineland the Good*, London, 1890. A facsimile edition of the whole Flatey Book was issued by Levin and Munksgaard, Copenhagen, 1930.

[6] In all probability these formed one continuous account, as Gustav Storm and others have maintained. He and others call them both the "Tale of the Greenlanders," and they will be so referred to here. Sometimes they have been called the "Saga of Eric the Red."

[7] *Aarbøger for Nord. Oldk. og Hist.*, 1887, pp. 305-313, and 1915, pp. 205-221.

"Leif put to sea and was tossed about on the ocean for a long time and came upon lands that he had no knowledge of before. There were fields of self-sown wheat and grapevines. There were those trees which are called 'mausur,'[8] and of all these they took specimens. Some of the timbers were so large that they were used in building. Leif came across some men in a shipwreck and brought them home with him. He showed manliness and generosity in this, as well as many other matters, such as in introducing Christianity into Greenland, and ever since he was called Leif the Lucky. Leif landed in Ericsfjord and went to his home in Brattahlid, where he was well received by all."

It appears from the story that Leif intended to take the shortest route between Norway and the Eastern Settlement in Greenland, both going and coming. Gustav Storm believes[9] that Leif was the first to attempt this direct route and for that reason was diverted from his course on his way to Norway and landed in the Hebrides, which lie somewhat farther south than the direct line between the southern point of Greenland and Nidaros. Subsequently a route was established between the Shetlands and the Faeroes.[10]

[8] The term "mausur" is generally considered to have referred to a species of maple.

[9] *Aarbøger for Nord. Oldk. og Hist.*, 1887, p. 313.

[10] See *Landnámabók*, Chapter 2.

FIG. 3—Norse ruins in Greenland. (Photograph by Fred Herz, Second Greenland Expedition of the University of Michigan, 1927.)

FIG. 4—Ericsfjord in the Eastern Settlement of Greenland (near the present Julianehaab): the outlook from Eric the Red's home in Brattahlid at the head of the fiord.

LEIF ERICSSON

And it may have been Leif's intention to follow this, but in all probability he lost his way through stormy weather.

Ever since the stories relating to these discoveries became generally known through the writings of Arngrim Jónsson the Learned[11] and of Thormód Torfason[12] many theories have been advanced as to the position of the countries where Leif found the plants mentioned above. The consensus of opinion seems to be that he landed on the east coast of North America, somewhere within the region of wild grapevines, that is not farther north than New Brunswick, nor farther south than Maryland. It seems most probable that he came ashore somewhere on the coast of New England (Fig. 5).

The Saga says explicitly that Leif saw more than one country. Possibly he came within sight of Nova Scotia and Newfoundland as he was driven southward by the storm. In all probability the gale was from the north and northeast and, therefore, the boat was not out of heavy seas until it passed the southern extremity of Nova Scotia. Afterward Leif probably sighted the shores of New England on the starboard side and sought to land there. The only

---

[11] *Gronlandia edur Grœnlands saga*, Skálholt, 1688.

[12] *Historia Vinlandiæ antiquæ*, etc., Copenhagen, 1705, and *Gronlandia antiqua*, etc., Copenhagen, 1706.

8   THE VINLAND VOYAGES

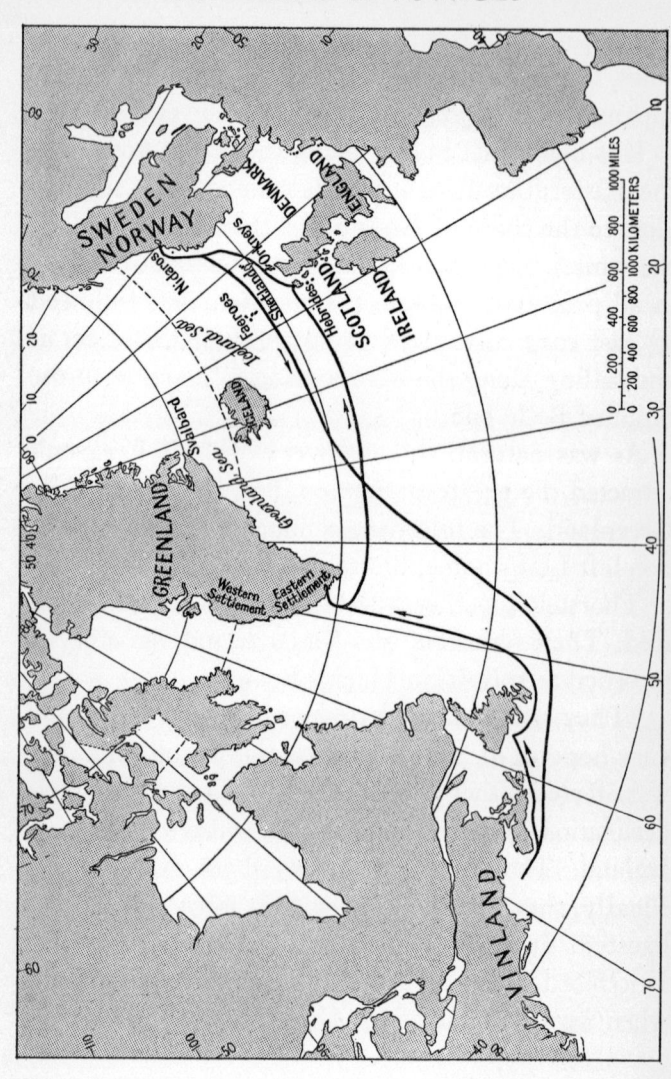

FIG. 5.—Map showing the conjectural route of Leif Ericsson's voyage in the year 1000. Scale, 1:50,000,000. (His route from Greenland to Norway in 999 is also shown.)

description of the country is the reference to the grape-vines, the self-sown wheat, and the "mausur" trees, apparently all common in New England vegetation.

It is likely that in continuing the return voyage to the Eastern Settlement in Greenland, Leif sailed so close to the shore of Nova Scotia that he saw land all the while; but, when he reached Newfoundland, he in all probability took to the open sea, and followed its east coast instead of going through Cabot Strait and sailing along the west coast until he came to the Strait of Belle Isle.

As was natural, this discovery of Leif Ericsson's attracted the greatest attention, first and foremost in Greenland. The following summer, 1001, an expedition left Leif's home, Brattahlid, headed by his brother Thorstein together with their aged father, Eric the Red. Their objective was Vinland, and the story is recorded as follows in Hauk's Book:

"They sailed out of Ericsfjord in good spirits and very hopeful of the adventure. For a long time they were tossed hither and thither without reaching their destination. They sighted Iceland and saw birds from Ireland. Their ship was driven about over the sea. Finally, they returned in the fall, worn out and exhausted. It was well-nigh winter when they reached Ericsfjord. Then Eric said: 'We were more cheerful when we sailed out of the fjord in the summer than we are now, yet it might have been worse.'" He had been very reluctant to join the expedition.

According to the Saga, this voyage was undertaken in a vessel that Thorbjörn Vifilsson, of Laugabrekka in Snæfellsness, had bought in Iceland the previous summer and emigrated in to Greenland. Therefore, they did not use Leif's ship, and there is no reference to Leif in connection with the above expedition. The inference might be that he had left previously in his own ship for Norway and that he had possibly not returned home for several years. He is not mentioned further in the Saga, although he lived some twenty years after his epoch-making voyage and died about 1020, then around fifty years of age. One is nevertheless justified in assuming that among the twenty men who accompanied Thorstein and Eric on their expedition were some who had been with Leif on his voyage the preceding summer; they presumably acted as guides. Unfortunately for this enterprise of Thorstein and Eric, they must have taken a too easterly direction. Contrary winds were doubtless responsible for the fact that they went so far to the northeast that they saw Iceland and then again so far to the southeast that they approached the Irish coast.

When Thorstein Ericsson returned to Greenland, he married Gudrid, only daughter of Thorbjörn, the owner of the ship on which the voyage of exploration

FIG. 6—Facsimile of the passage in Hauk's Book containing the story of Leif Ericsson's finding of Vinland. (Arna-Magnaean codex No. 544, 4to, back of leaf 96, lines 16-33.)

FIG. 7—Lysufjord in the Western Settlement of Greenland (near the present Godthaab).

THORFINN KARLSEFNI

had been made. Thorstein and his wife set up a home at Lysufjord in the Western Settlement of Greenland. During the latter part of the winter Thorstein became ill and died. Shortly afterward Gudrid's father, Thorbjörn, succumbed to disease, and thereafter Gudrid moved to Brattahlid and made her home with her father-in-law, Eric the Red.

No attempt was made at a Vinland voyage during the summer of 1002. In the fall two ships came from Iceland. One belonged to Thorfinn Thórdarson, called Karlsefni, a member of a distinguished family in Skagafjord in the north of Iceland. One of Thorfinn's passengers was Snorri Thorbrandsson from Alptafjord, on the west coast of Iceland, otherwise known from the Saga of the Ere-Dwellers Eyrbyggja Saga). The other ship was commanded by Bjarni Grimulfsson from Breidafjord and Thorhall Gamlason from eastern Iceland. Eric invited the owners of these ships and their crews to spend the winter with him in Brattahlid. Just after Christmas, Thorfinn and Eric's daughter-in-law, Gudrid, were married. Thereafter the Saga concerns itself chiefly with Thorfinn (Chapters 7-14); hence it has sometimes been called the Saga of Thorfinn Karlsefni. In fact, the major part of the Saga deals with him and Gudrid, rather than with Eric and his sons. Chapters 1, 3, 4, and 6 relate Gudrid's story, and Chapters 7 to 14 concern themselves with Thorfinn. Chapter 1 may also be said to deal with Thorfinn inasmuch as it gives a genealogy of his ancestors taken from the Icelandic Book of Settlement (*Landnámabók*). The author, curiously, has failed to connect the subject matter of Chapters 1 and 7.

In the spring of 1003 Thorfinn and Snorri, Bjarni and Thorhall set out in their ships on a voyage of discovery directed towards Vinland, and on Thorbjörn's ship went many people of Greenland. In the beginning of the narrative dealing with this voyage the new land is for the first time called Vinland (the Good) and the expedition a Vinland voyage. Nowhere previously, neither in the narrative of Leif's discovery nor in that of the search of Thorstein and Eric for the country, is the land characterized as Vinland. It would appear from the Eric Saga that Leif did not name the country, but that it received its name afterwards in Greenland, where it was much discussed. Whether it was called "the Good" from the beginning or whether that epithet was applied later is uncertain.

It is now in order to quote the various portions of the narrative about this voyage from Hauk's Book and analyze each in turn. The beginning is as follows:

"Vinland the Good and the desirability of going there became a popular topic of discussion at Brattahlid, and it was said that the new land had many advantages. Finally Thorfinn Karlsefni and Snorri made preparations to go there in the spring. With them went also Bjarni and Thorhall, who were mentioned above, in their own ship. There was a man named Thorward. He was married to Freydis, the natural daughter of Eric the Red. He went with them, and also Thorhall who was called the Hunter. He had been with Eric the Red a long time, his hunter in summer, and his steward in winter. He was a big man and strong, swarthy and giant-like, a man of few words but with a wicked tongue and had a reputation for inciting Eric to evil. He was a poor Christian. His knowledge of the unsettled regions was extensive. On the voyage he was on the same ship with Thorward and Thorwald. They used the same ship in which Thorbjörn had come to Greenland. There were in all 160 people who sailed to the Western Settlement in Greenland, thence to Bear Island. From there they sailed to the southward two 'dægr.'[13] Then they sighted land, launched a boat, and found there large flat stones some of them twelve ells wide. There was also a number of Arctic foxes. They gave a name to the land and called it Helluland (i.e. Flatstoneland). Then they sailed south-southeast two 'dægr'

[13] "Dægr," according to the Norse way of recording time, was usually twelve hours.

and came upon a wooded land, with many wild beasts. An island lay to the southeast; there they slew a bear and named the island Bear Island, but the land they called Markland (i.e. Forestland). Then they sailed south along the shore for a long time and finally came to a cape. The land lay to the starboard; there were long strands and sandy banks. They rowed to the land and found there the keel of a ship and thence called the cape Keelness. The strands they also called Furdustrandir (Wonderstrands), because it took so long to sail past them. Then the land became indented with bays.[14] They sailed their ship into a bay; there was an island at the mouth of the bay, and there were strong currents around it. They called the island Straumey (Stream Island). There were so many eider ducks on the island that it was scarcely possible to walk because of the eggs. They named the fjord Straumfjord (Streamfjord). Then they carried their cargoes ashore and settled there. They had brought all kinds of livestock with them. The country was beautiful, and they occupied themselves only with exploring it. They spent the winter there but had not provided for it during the summer. In time fishing failed, and it became difficult to obtain food.[15] Then

---

[14] Here a portion is omitted referring to Haki and Hækja, scarcely original in this connection. It will be referred to later (p. 26).

[15] The story of Thorhall the Hunter and the whale is omitted here.

they prayed to God for help. The weather thereupon improved, and they could now row out and fish; and thereafter there was no lack of provisions, there was game on the land, eggs on the island, and fish in the sea."

One draws the conclusion from the narrative that the voyage was essentially one of colonization; that is why so many went there, 160 men and women, and carried livestock with them. However, it is not explained why Thorfinn went first north to the Western Settlement, before leaving Greenland. Some have conjectured that it had something to do with the property that Gudrid inherited there through Thorstein, her first husband. It has also been suggested that Thorfinn and his men preferred to leave from the Western Settlement because with a northerly wind they could take a more westerly course than if they set out from the Eastern Settlement.[16] The location of Bear Island, to which Thorfinn and his men sailed from the Western Settlement, is an open question. An ancient description of Greenland declares that it lies towards the north, nine days' rowing from Lysufjord; further, that it takes twelve days to row around it. In view of this, it would appear that the island referred to in the Saga is the island of Disko, although,

[16] H. P. Steensby, *The Norsemen's Route from Greenland to Wineland*, Copenhagen, 1918, pp. 33ff.

on the other hand, it seems unlikely that Thorfinn and his men went there. Yet there is hardly any other island than Disko that can claim consideration. Of course, there may be a misstatement in the Saga with reference to this journey so far north; or there may have been some reason for it omitted by the sagaman, either through oversight or ignorance.[17]

It is stated that Thorfinn came upon Helluland after two days' sailing from Bear Island, something that could not be true even if Bear Island were directly out from Lysufjord, as Gustav Storm believed. If Thorfinn and his men did not sail directly south but took a southwesterly course, the shortest way to the next land, the distance is so great that they could not have covered it in two "dægr," even though two "dægr" signifies two days, as it probably does.[18] It has been suggested by Finnur Jónsson that there is a mistake in the manuscript, the writer having written "ii" for "u," that is v, or 5, which is possible.[19] Yet it seems to me more likely that the scribe may have mixed his sentences and taken this number from the voyage to Markland which follows.

[17] It is possible that there were plentiful provisions in this Bear Island and that the voyagers went thither to get supplies before starting for the unknown lands.

[18] *Grønlands historiske Mindesmærker*, Vol. 1, pp. 162-163.

[19] *Historisk Tidsskrift* (Norwegian), Ser. 5, Vol. 1, Christiania, 1912, p. 133.

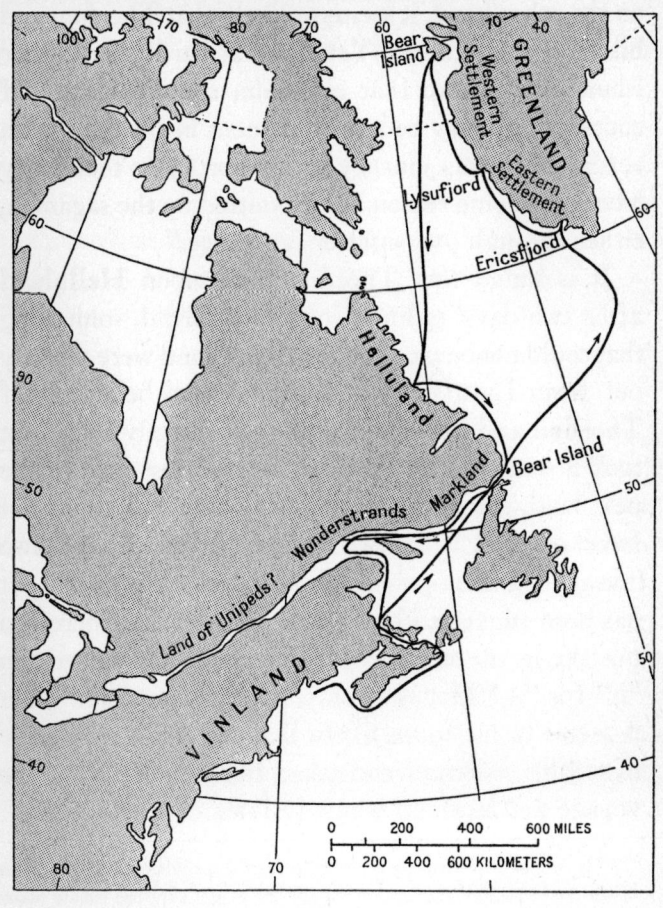

FIG. 8—Map showing the conjectural route of Thorfinn Karlsefni's voyage, 1003-1006(?). Scale, 1:32,000,000.

It is generally conceded that Helluland (Flatstone-land) was the northeast coast of Labrador. No one, however, has to my knowledge pointed out a definite place particularly characterized by flat stones, but Arctic foxes are common there.[20]

There is a difference of opinion concerning the identification of Markland. Gustav Storm was inclined to think that Thorfinn had gone to Newfoundland, basing his opinion on the sentence stating that he took a more southeasterly direction from Helluland, which corresponds with the position of the two places under consideration. Because of the coast line, Thorfinn was obliged to take a more easterly course after leaving Labrador. Storm believed that Newfoundland was Markland, which is quite possible although the description of the voyage fits the eastern extremity of Labrador better, as is pointed out by H. P. Steensby in his work on the route followed by Thorfinn.[21] The description of Markland is somewhat sketchy in the Saga, but as far as it goes it might easily fit the southeastern coast of Labrador, for there are great forests there, especially a short distance from the shore. In the Tale of the Greenlanders the description is somewhat more detailed: "The land was heavily wooded, and white sands sloped down towards

[20] *Aarbøger for Nord. Oldk. og Hist.*, 1887, p. 335.
[21] H. P. Steensby, *op. cit.*, pp. 38f.

FIGS. 9 and 10—Coast of Labrador near Cape Mugford, in latitude 58°N. (Photographs by Putnam Baffin Island Expedition, 1927 (Fig. 9) and D. B. MacMillan expedition to Labrador, 1911 (Fig. 10).)

the sea." Such a description is said to apply well to the southeastern portion of Labrador. On the north side of the Strait of Belle Isle is a place called Blanc Sablon (White Sand) because it is said to appear white from the sea in the sunshine.[22]

Steensby considered that the Bear Island lying southeast of Markland is the northernmost point of Newfoundland. This is not necessarily the case, for the reference may be to Belle Isle, which lies about 10 kilometers to the north from the shore. Steensby thought that Thorfinn had sailed along the mainland and thus reached the strait. The locality is usually foggy; therefore it is unlikely that they noticed Newfoundland, still more unlikely that they considered it an island in the event that they went ashore and killed a bear.

Then follows the narrative of the voyage after leaving Markland (Fig. 11). It is somewhat indefinite, particularly as to the location of Keelness. It is certain, however, that they sailed a long time with land on the starboard, also that the shore line was long and sandy.

[22] Professor Halldór Hermannsson pointed this out in an article entitled "The Wineland Voyages," *Geogr. Rev.*, Vol. 17, 1927, pp. 107-114. Previous to that Hermannsson had written a long article about the same subject in *Tímarit Thjódrœknisfélags Islendinga*, Winnipeg, Vol. 1, 1919, pp. 25-65. He deals there with the earliest records, as well as the various efforts at a solution of the problem.

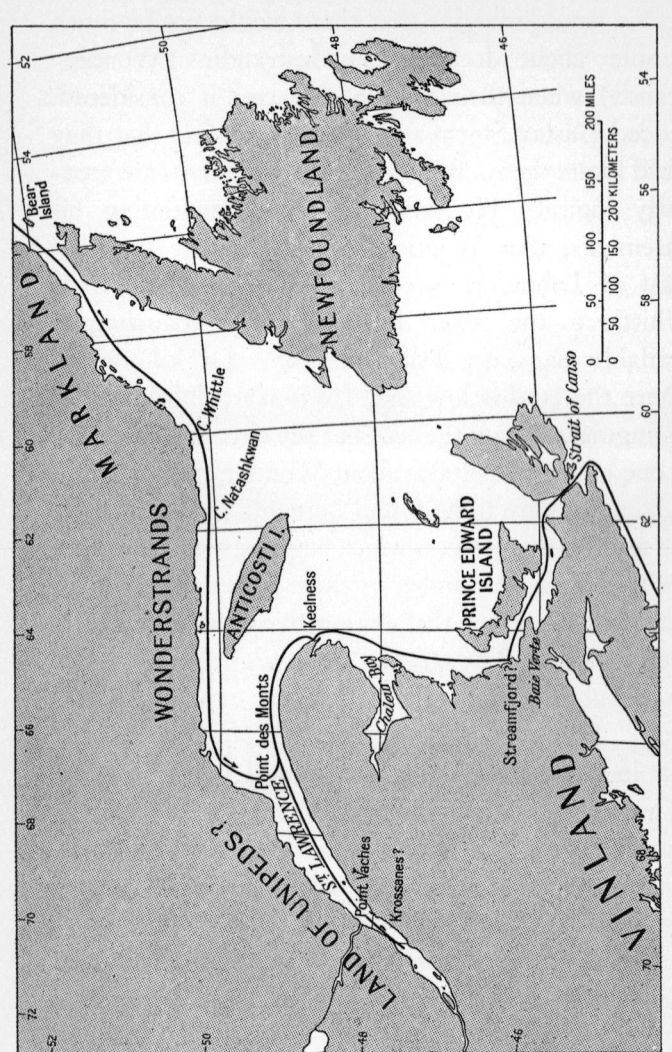

FIG. 11—Map showing in detail the conjectural route of Thorfinn Karlsefni in the Gulf of St. Lawrence. Scale, 1:10,000,000. The route in the St. Lawrence to Krossanes (Crossness) represents the search for Thorhall the Hunter.

One would imagine that there would not be much trouble about locating Furdustrandir (Wonderstrands) when their unique character is considered. Indeed Gustav Storm and others have felt that they could locate them. But Steensby's arguments are especially logical. He pointed out, substantiating his statements, that Wonderstrands were the southern coast of Labrador, especially the shore from Cape Whittle to the Seven Islands, which, according to available maps, is a distance of some 440 kilometers. There the land is low and, from afar, difficult to be distinguished from the sea; but the coast is harborless, as one of the texts states about Wonderstrands. Slight indentations are to be found here and there, and into the sea flow many rivers which have gathered large deposits of sand. A number of these rivers contain magnetic iron; and for that reason these desolate shores are inhabited. The strands are straight from east to west, but the Saga relates that the explorers sailed south. However, that may be explained by the fact that in going from Belle Isle (Bear Island?) to Cape Whittle they sailed southwest about 330 kilometers. Furthermore, it is well known that the ancient Icelanders were not as accurate about directions as men are nowadays, particularly at sea.

Halldór Hermannsson is of the opinion that Leif Ericsson went through the Gulf of St. Lawrence on his way back to Greenland from Vinland and that he passed through the Strait of Belle Isle.[23] No doubt some of Leif's men were with Thorfinn and influenced his choice of a course. It seems to me likely that Leif in following such a course would have been aware that he had Newfoundland to the starboard on his way north; and in that case those of his men who in all probability were with Thorfinn would have recommended sailing south along the coast of Newfoundland. But I believe it is more probable that Leif had Nova Scotia and Newfoundland to the larboard on his way back from Vinland; and that may have influenced Thorfinn and his men to sail southwest and west from Labrador (Markland and Wonderstrands) with land to the starboard. In this way Leif's voyage north influenced Thorfinn's voyage south, as Halldór Hermannsson suggests. The course was somewhat similar. Thorfinn and his companions may have considered themselves to be following the same route as was followed by Leif on his voyage home, that is until they came to the estuary of the St. Lawrence; then they found that they had not the open sea to the larboard.

Then comes the question of the location of Keelness, which is difficult to determine from the refer-

[23] Hermannsson, Wineland Voyages, *Geogr. Rev.*, Vol. 17, pp. 108-109.

FIG. 12—Western coast of Newfoundland at the Bay of Islands.
FIG. 13—Northeastern coast of Newfoundland: Notre Dame Bay.

ences in the Saga. The inference is that Thorfinn and his men had already passed Wonderstrands when they set out in their boats and landed at Keelness; and by that time the shore had become indented with bays. One can better understand the narrative if the sentences are arranged thus: "Then they took a southerly course along the shore for a long time with land to starboard. The shore line was long and sandy. They called the strands Wonderstrands, because they sailed past them for such a long time. They came to a cape, rowed ashore, and there found the keel of a ship. Therefore they called the cape Keelness." Steensby believed that Keelness was Point Vaches, located on the northern side of the Saguenay River mouth, rather than Point des Monts, the extremity of a wide cape in the St. Lawrence, about 200 kilometers farther to the northeast. In reality, Point Vaches is scarcely a cape; and it is so far up the St. Lawrence that it is highly improbable that it is Keelness. It is more probable that Keelness was Point des Monts, especially as it is unlikely that Thorfinn and his men sailed farther west or into the St. Lawrence. The estuary is not so wide at this point that the land to the south is not in plain sight. Furthermore, the narrative does not state that Wonderstrands had been passed when Thorfinn and his companions landed at Keelness. From the story of Thorhall the Hunter's return voyage (see p.

32) one may perhaps infer that Keelness was not west of Wonderstrands. The story relates: "They sailed north past Wonderstrands and Keelness and sought a westerly course."[24] From that description alone one might consider that Keelness was either Cape Whittle or Point Natashkwan. A further light is thrown on Keelness later in the story: "Thorfinn went in one ship to seek Thorhall the Hunter; the rest of his party remained behind. They sailed to the northward around Keelness and then bore to the westward with land to the larboard. The country was a wooded wilderness as far as eye could reach, not an open space anywhere. When they had sailed a long time, they came to a river that flowed out of the land from east to west. They entered the mouth of the river and made for the southern bank."[25] Then the Saga relates the shooting of Thorwald Ericsson and continues: "They then departed towards the north and were under the impression that they sighted the Land of the Unipeds. They did not risk going any farther. They believed that the mountains that they saw there were the same as those they saw in Hóp, being at an equal distance both ways from Streamfjord." On two occasions, when they went to Hóp and when they went to seek Thorhall the Hunter, they sailed from

[24] *Hauksbók*, Copenhagen, 1892-96, p. 439.
[25] *Ibid.*, p. 442.

FIGS. 14 and 15—Cape Gaspé at the end of the Gaspé Peninsula: above, the extreme point of the cape and, below, a secondary headland, Cape Bon Ami, back of the main point. (Photographs from Geological Survey of Canada.)

Streamfjord. The story about the first departure is as follows: "To turn to Thorfinn Karlsefni, he went southward off the land, and Snorri and Bjarni and their men with him. They sailed for a long time, until they came to a river that flowed from inland into the sea. There were great bars at the mouth of the river so that it could only be entered at flood tide. Thorfinn and his men sailed into the mouth of the river and called it Hóp."[26]

If one examines the map, it becomes apparent that the references to Streamfjord, Hóp, Keelness, etc., do not fit Steensby's theory either that Keelness is Point Vaches or that Hóp is Montmagny, far in the estuary of the St. Lawrence. Everything seems to support the argument advanced by Halldór Hermannsson[27] that Streamfjord is on the eastern coast of New Brunswick and that Keelness is Cape Gaspé, the extremity of Gaspé Peninsula (Fig. 11). Assuming that, one understands why Thorfinn sailed southward round the land and arrived in Hóp, which must have been to the southeast of New Brunswick. And then, also, one realizes why he later sailed northward around Keelness and there bore to the westward with land to the larboard. It is thus possible to say that he took a north-

[26] A small land-locked bay.
[27] Hermannsson, Wineland Voyages, *Geogr. Rev.*, Vol. 17, pp. 110-111.

erly course around the peninsula, both coming and going, to Streamfjord. A reference will be made to this later.

The Saga relates that after Thorfinn and his men had left Wonderstrands the land became indented. There is a bay north of the Seven Islands but no other bays until one reaches Point des Monts. But, upon arriving there and even before in clear weather, one sees the land across the St. Lawrence, especially the Gaspé Peninsula. It is possible that Thorfinn and his men, not caring to follow the desolate Wonderstrands any longer, turned south towards the Gaspé Peninsula. On the eastern coast of the peninsula are the Bay of Gaspé and Malbaie.

According to the story, Thorfinn and his men sailed into a bay; and then follows the episode about Haki and Hækja. As Finnur Jónsson has pointed out, that episode does not seem to fit in here but is probably an insertion taken from the narrative of Leif's voyage.[28] It may have a historical basis, but it can be omitted here without causing any interruption in the story about Thorfinn's voyage, which now goes on to tell about Streamfjord.

There has been much difference of opinion as to the location of Streamfjord. Assuming, as I do with

[28] *Historisk Tidsskrift* (Norwegian), Ser. 5, Vol. 1, 1912, p. 28; cf. *Aarbøger for Nord. Oldk. og Hist.*, 1915, p. 20.

THORFINN KARLSEFNI

Steensby, that Wonderstrands were the southern coast of Labrador, it is not necessary to discuss the theories of those who think that Thorfinn sailed east of Newfoundland. Steensby[29] maintained that the estuary of the St. Lawrence was Streamfjord and Hare Island was Stream Island. He also believed that the place named Hóp by Thorfinn and his men was Montmagny (St. Thomas) by the Rivière du Sud, not so very far below the city of Quebec on the other side of the St. Lawrence. Gathorne-Hardy, in his book on the Norse discoveries of America, disagrees with Steensby's theories,[30] but Gustav Holm has endeavored to refute his objections and has defended Steensby's views as at least possible.[31]

The fact that the Saga mentions numerous indentations is worthy of consideration, for if the explorers sailed with land to the starboard after leaving Wonderstrands, past Hare Island, they would not have noticed many indentations. It is said that Stream Island lay outside, as they sailed into the bay, which does not correspond with the location of Hare Island, which lies in the St. Lawrence, about 235 kilometers

[29] Steensby, *Norsemen's Route*, pp. 53ff.

[30] G. M. Gathorne-Hardy, *The Norse Discoverers of America*, Oxford, 1921, pp. 241-243.

[31] Gustav Holm, Small Additions to the Vinland Problem, *Meddelelser om Grønland*, Vol. 59, 1925, pp. 11-37; reference on pp. 31-34.

above Point des Monts; nor is it the only island, for there is Green Island, too, and Pilgrim Islands, somewhat farther in, and many other islands beyond.

Furthermore, with Steensby's identification of Streamfjord as the St. Lawrence, there could not be any justification for the phrase used in the Saga about Thorfinn's sailing "northward around" the land in search of Thorhall, although "southward off the coast" in case of his sailing for Hóp might find application.

Halldór Hermannsson suggests that Streamfjord is Chaleur Bay, south of the Gaspé Peninsula.[32] There is an island lying outside known as Miscou Island; in fact there is another island there called Shippigan Island, possibly not mentioned in the Saga because it is almost connected with the former. It is not unlikely that the voyagers considered the latter a peninsula, because the strait that separates it from the mainland is so narrow. There probably are currents at high tide around Miscou Island; in any case there is a difference of something like ten feet between high and low tide at the head of the bay. Miscou Island is rich in bird life at present, although there are probably not as many eider ducks and eggs there now as the Saga relates; but that could hardly be expected. The Saga

[32] Hermannsson, Wineland Voyages, *Geogr. Rev.*, Vol. 17, pp. 110-112.

Fig. 16—Head of Chaleur Bay. (Photograph from Geological Survey of Canada.)

Fig. 17—Bird ledges, Bonaventure Island, Gaspé Peninsula. (Photograph by P. A. Taverner.)

relates that the scenery was beautiful in Streamfjord. Halldór Hermannsson refers to Cartier's description of Chaleur Bay. His description, written after his voyage in 1534, is as follows: "The land toward the south of the said bay is as fair as good land, arable, and full of as goodly fields and meadows as we have seen, and level as a pond; and that toward the north is a high land, with mountains all full of trees of tall growth, of many sorts, and among others are many cedars and firs, as goodly as it may be possible to behold for to make masts sufficient to mast ships of three hundred tons or more; in which we did not see a single place devoid of woods, save in two places of low lands, where were very beautiful meadows and lakes." Then Cartier continues in his praise of the climate, plant life, wild grain, various berries, saying that the country is covered with red roses and other fragrant flowers and that all the lakes are full of salmon. In this connection it is well to note that the locality is in latitude 48° N.; that is one degree farther south than Paris. The winter is, however, no less severe than that which Thorfinn and his men experienced in Streamfjord; excellent fishing is also to be found there today. When Cartier arrived five centuries ago the Indians did extensive fishing there. Grapevines do not, however, grow there nor anywhere else in New Brunswick, a fact that agrees well with the Saga,

Thorfinn's search for the land farther south, and Thorhall the Hunter's dissatisfaction and departure, as will be pointed out presently.

Halldór Hermannsson's conjecture that Streamfjord is Chaleur Bay is in some ways very reasonable. Considering, however, the description in Chapter 8 of the Saga and the account in Chapter 9 of the preparations for the departure of Thorhall the Hunter, it would seem that Streamfjord was not as large as Chaleur Bay. The main bay is 136 kilometers long from Miscou Island and 25 kilometers wide. It terminates in another smaller bay, which is 35 kilometers in length, and outside of which there also is an island, called Heron Island. Towards the south and north there extend two other large bays, besides a small one towards the south called Caraquet Bay, at the mouth of which lies an island of the same name. The difficulty lies in identifying Chaleur Bay with Streamfjord and at the same time assuming that Cape Gaspé is Keelness, nor does it correspond with the narrative in Chapter 12 of the Saga dealing with Thorfinn's search for Thorhall the Hunter. Yet nothing in the narrative contradicts the assumption that Streamfjord is to be found on the east coast of New Brunswick.

Thorfinn and his men evidently soon came to the conclusion that they had not found the Vinland de-

Fig. 18.—The cape at Percé, Gaspé Peninsula, showing Percé Rock and Bonaventure Island. (Photograph from Geological Survey of Canada.)

scribed by Leif Ericsson; they did not discover any grapes in the locality where they wintered. In all probability they explored the country in the summer of 1003, and they must have seen the St. Lawrence from the Gaspé Peninsula, particularly if Streamfjord is Chaleur Bay.

In the spring of 1004 Thorfinn headed one expedition to seek Vinland, and Thorhall the Hunter headed another. The story of this commences thus: "It is told that Thorhall wished to go north of Wonderstrands and thus to seek Vinland, but Thorfinn decided on a course leading southward around the country." Presumably the Saga is not altogether right here about Thorhall's intentions; yet he may have pretended that he wished to seek Vinland, although it is difficult to see how he expected to find it by going north of Wonderstrands. The account of his voyage and the ditties ascribed to him, especially the second, show that he had in mind to go back to Greenland. About his voyage the following is told: "Thorhall made the preparations out near the island for his trip; there were nine men altogether, but all the others went with Karlsefni; and when Thorhall carried water to his ship and drank of it he recited this ditty:[33]

---

[33] The translation of the two ditties is by A. M. Reeves, *op. cit.*, pp. 45-46.

'When I came, these brave men told me,
　Here the best of drink I'd get,
Now with water-pail behold me,—
　Wine and I are strangers yet.
Stooping at the spring, I've tested
　All the wine this land affords;
Of its vaunted charms divested,
　Poor indeed are its rewards.'

"And when they were ready they hoisted sail. Then Thorhall recited this ditty:

'Comrades, let us now be faring
　Homeward to our own again!
Let us try the sea-steed's daring,
　Give the chafing courser rein.
Those who will may bide in quiet,
　Let them praise their chosen land,
Feasting on a whale-steak diet,
　In their home by Wonder-strand.'

"Thereafter they sailed northward past Wonderstrands and Keelness, intending to turn to the westward. But they encountered westerly winds and were driven to Ireland and were maltreated there and thrown into slavery. There Thorhall lost his life, according to what traders have told."

THORFINN KARLSEFNI

The ditties of Thorhall are most important; in fact, they are the oldest sources we possess about these events. In the first ditty it is to be noted that he complains that he has not tasted a drop of wine, and in the second that he urges his men to go home. His declaration that he was going to seek Vinland was only a blind. He merely pretended that he was going to follow the Gaspé Peninsula towards the southwest in search of Vinland; and from the account of Thorfinn's voyage in search of Thorhall it appears that the former knew of that plan and therefore went in that direction, as Halldór Hermannsson has suggested and as I think is right. It is possible that Thorhall and others considered the St. Lawrence a narrow strait through which one could go and reach Vinland. If he had sailed far enough up the St. Lawrence, he would also have found grapevines. Steensby considers that Leif Ericsson's Vinland was to be sought there, as well as Thorfinn's Hóp, where he and his men found the grapevines, in support of which he points out that Cartier found such quantities of grapes on the Isle d'Orléans that he named it Isle de Bacchus.[34] The last part of the narrative about Thorhall, namely their slavery in Ireland and reports of it by Norse traders, is not at all unlikely.

As to the narrative dealing with Thorfinn and his

[34] See Steensby, *Norsemen's Route*, pp. 72ff.

voyage, it is well to consider the beginning and the end of it in order to trace the course he followed:[35]

"Now is to be told of Karlsefni that he went southward around the country, and Snorri and Bjarni with their people. They sailed a long time until they came to a river which flowed down from the land into a lake and so into the sea. At the mouth of the river were great bars so that it could be entered only at the height of the flood tide. Thorfinn and his men sailed into the mouth of the river and called it Hóp. In the hollows they found self-sown wheatfields and on the hills grapes. Every brook was teeming with fish. They dug pits on the shore where the tide rose highest, and when the tide fell there were halibut in the pits. Wild animals of all kinds were roaming in the woods. They remained there a fortnight and enjoyed themselves without any disturbance. They had their livestock with them. One morning, however, when they looked about they saw a large number of skin canoes approaching." The Saga then continues and tells about the encounter between Thorfinn's party and the Skrælings. From the account it is evident that the Skrælings did not come from the interior but from the sea. They left peacefully the first time and rowed south around the cape. Then the Saga goes on:

"Karlsefni and his men had built their huts above

[35] *Hauksbók*, pp. 439-442.

FIG. 19—Facsimile of the page in Hauk's Book containing the description of Hóp. (Arna-Magnaean codex No. 544, 4to, front of leaf 100.)

the lake, some nearer, others farther away. There they spent the winter. There fell no snow on the ground, and their livestock lived by grazing. At the beginning of spring they discovered one morning a large number of skin canoes rowing from the south past the cape. It looked as if coals had been broadcast over the bay." Karlsefni and his men started bartering with the strangers. All went well until a bull, belonging to Karlsefni, came bellowing from the woods. This so frightened the Skrælings that they ran to their canoes and rowed southward. Then nothing happened for three weeks. At the end of that time the Skrælings came down in a multitude of boats from the south like a torrent. Then the Skrælings attacked Karlsefni and his men "who had to flee up along the river, because it appeared as if the Skrælings came from all directions, and they did not halt until they reached some crags and resisted there with all their might." The Saga then gives an account of Freydis and the flight of the Skrælings, closing thus: "It now appeared to Karlsefni and his men that, although the country had good resources, there would constantly be danger and war brought on by the aborigines. They prepared to leave the place, intending to go back to their country, and sailed northward around the land. . . . Later Karlsefni and his men discovered a cape upon which was a large number of animals, so that it looked as if it were a cake of dung, because of the animals that lay there at night. Karlsefni and his men arrived in Streamfjord and found there plenty of everything that they needed." The author adds the following remark: "Some say that Bjarni and Gudrid and one hundred men with them remained behind (in Streamfjord) and did not go farther, but Thorfinn and Snorri and forty men with them went south, staying in Hóp barely two months and returning the same summer." Then follows the account of Karlsefni's trip to seek Thorhall the Hunter, which has been dealt with above (p. 24). In this connection it is worth noticing that that voyage rather indicates, as Gustav Holm has pointed out,[36] that Karlsefni did not spend a whole winter in Hóp but returned the same summer to Streamfjord. Yet the description of their voyage to Hóp and all that happened to them there may be historical nevertheless.

It was mentioned above that Keelness was possibly Cape Gaspé, and Streamfjord Chaleur Bay, as Halldór Hermannsson suggests. Assuming this, one must conclude that, when it is said that Karlsefni sailed from Streamfjord "southward round the country" to Hóp, he went south around Nova Scotia. Now the story relates in connection with their voyage north when seeking Thorhall the Hunter that "They sailed

[36] Holm, *op. cit.*, pp. 33-34.

north around Keelness, and bore along to the west, and that they thought that the mountains in Hóp and those that they now found were the same, and that they were about equal distance removed from Streamfjord in either direction." This does not make it probable that Streamfjord was Chaleur Bay, for it is farther east, south, and then west around Nova Scotia out of Chaleur Bay than east, north, and then west around Gaspé Peninsula, even if one travels well up the estuary of the St. Lawrence. On the other hand, it is possible that they shortened the distance around the eastern coast of Nova Scotia by sailing south of Prince Edward Island and then through the Strait of Canso, between Cape Breton and the mainland. This is perfectly possible, although fishermen who centuries later fished around Newfoundland did not know about this strait for a long time.[37] But if the Saga is right as to the equal distances from Streamfjord in case of these two voyages, the theory that Streamfjord is Chaleur Bay is hardly tenable, and we must therefore look for it somewhere farther south on the coast of New Brunswick. And apparently only three bays along that shore could come in for consideration, that is Miramichi Bay, Cocagne Harbor, and Shediac Bay; but which of these would be

[37] *Aarbøger for Nord. Oldk. og Hist.*, 1887, p. 333.

most likely to fit the description of Streamfjord I am not in a position to say.

The question arises as to the real location of the place that Thorfinn and his followers called Hóp. Most writers on the subject agree that it was somewhere on the coast of New England at the mouth of some river there. Gustav Storm, in his article written in 1887,[38] maintained that Nova Scotia was Vinland and that Hóp was on the southeastern coast of that province. He felt that he had definite proof that grapevines and self-sown wheat had been found there when the Europeans first came to that locality and settled there in the seventeenth century.[39] Many were inclined to agree with Storm in regard to this, as well as to other theories advanced by him on the subject. In 1910 M. L. Fernald published a treatise[40] on the plants that had been found in Vinland and Hóp, maintaining that those particular species of plants, self-sown wheat and grapevines, were not found in Nova Scotia, as far as botanists have been able to ascertain. The effect of Fernald's criticism was that the theories of earlier writers, like Rafn, were revived and Vinland and Hóp sought farther south, on

[38] *Aarbøger for Nord. Oldk. og Hist.*, 1887, p. 334.

[39] *Ibid.*, pp. 340-342.

[40] M. L. Fernald, Notes on the Plants of Wineland the Good, *Rhodora*, Vol. 12, 1910, pp. 17-38. [See note, however, on p. 67, below. —H. H.]

THORFINN KARLSEFNI

the coast of New England. Steensby, however, contended that Leif's Vinland and Karlsefni's Hóp were far up the St. Lawrence. Gustav Holm agreed with Steensby that Hóp was there, but he believed that Leif's Vinland was on the southeastern coast of New England. Correctly he pointed out that there is nothing in the Saga of Eric the Red indicating that Leif and Karlsefni reached the same place.[41] But this has generally been assumed because the landfall of both had the same characteristics, that is, self-sown wheatfields and grapevines.

Gathorne-Hardy came to the conclusion that Streamfjord was the eastern extremity of Long Island Sound and that Hóp was the estuary of the Hudson River, in other words, New York Bay.[42]

Halldór Hermannsson does not attempt to fix the locality of Hóp, but he contends that it and Vinland must have been south of the northern limits of the wild grape, that is Passamaquoddy Bay,[43] which is on the boundary of New Brunswick and Maine. On a map that accompanies his article he places the name Vinland with an interrogation mark at the southern extremity of New England.

Gathorne-Hardy points out that various features of the earliest descriptions of the upper part of New

[41] Holm, *op. cit.*, pp. 32-37.
[42] Gathorne-Hardy, *op. cit.*, p. 276.
[43] Hermannsson, Wineland Voyages, *Geogr. Rev.*, Vol. 17, p. 112.

York Bay and the mouth of Hudson River correspond to the description of Hóp. The fallacy in his argument seems to me to be, that New York Bay and Hudson River are too far south, and even west, of the southwestern extremity of Nova Scotia. The distance from any of the bays on the east coast of New Brunswick into the St. Lawrence is much shorter than that from those bays all the way south to Long Island. Besides, although certain apparent similarity might be found between Hóp and New York Bay, there are other reasons why such a theory is highly problematic, and the comparison between the two is difficult at the present day, among other things, because probably the shore line around New York has been sinking during the last centuries.[44]

Before Gathorne-Hardy had developed his theory that Hóp was New York, several scholars had suggested other places. E. N. Horsford and his daughter, Cornelia Horsford, considered that Hóp was Back Bay, Boston, at the mouth of the Charles River. They believed that they had found Norse ruins at Cambridge and elsewhere. Cornelia Horsford engaged two noted Icelandic scholars, Thorsteinn Erlingsson and Valtýr Gudmundsson, to examine these ruins in the summer of 1896. They concluded

[44] W. H. Babcock, Early Norse Visits to North America, *Smithsonian Misc. Colls.*, Vol. 59, No. 19, Washington, 1913, p. 137.

that the remains resembled Icelandic ruins, but they found in them pieces of pottery assuredly from a later period and therefore left it to American experts to decide whether they were not remains of later settlements.[45]

William H. Babcock has written a valuable book on the Vinland voyages.[46] He draws attention to the fact that since Thorfinn's time the land has sunk a great deal on the eastern coast, and from that he concludes that the bay that E. N. Horsford and his daughter refer to was not there a thousand years ago.[47] He believes that Hóp was most likely what is now called Mount Hope Bay in the state of Rhode Island. This bay is landlocked in a way which the Icelanders styled "hóp"; the Taunton River drains into it. Two estuaries lead to the sea: one, the Sakonnet River, east of Rhode Island, which Babcock thinks may have been formed in recent centuries through the sinking of the land, and the other, the Bristol Narrows, to the northwest. The Bristol Narrows have an outlet into the Eastern Channel and Narragansett Bay. Formerly, bow-shaped sand bars had been across the channel where now are shallows. Babcock draws attention to other local features that correspond with the descrip-

[45] Cornelia Horsford, Vinland and Its Ruins, *Popular Science Monthly*, Vol. 56, 1899-1900, pp. 160-176.

[46] Babcock, *op. cit*.

[47] *Ibid.*, p. 136.

tion in the Saga, such as crags and capes. C. C. Rafn advanced the same ideas in his great work, *Antiquitates Americanae*, published in Copenhagen in 1837. The similarity of the name Hóp and Hope is interesting, the bay taking its name from Mount Hope near by. The pronunciation of the two names in English and Icelandic is the same. How did the hill and the bay receive that name? W. H. Munro, in his history of Bristol,[48] has suggested the possibility that some of Thorfinn's men had remained behind and taught the name to the Indians. It would be difficult to prove that; yet the origin of the name has not been satisfactorily explained.

Babcock was not certain that Hóp was Mount Hope Bay; he was almost as ready to believe that it was farther north—south of Maine, however. It is rather unlikely that Thorfinn and his men went south of Cape Cod. It is not improbable that Hóp was somewhere on the coast between Cape Cod and the northernmost point of the coast of Maine, but no definite place there has been plausibly suggested.

Steensby seems to have much to support the statement that Thorfinn and his men followed the coast after having discovered Helluland. When they sailed

---

[48] W. H. Munro, *Tales of an Old Sea Port: A General Sketch of the History of Bristol, Rhode Island, Including . . . an Account of the Voyages of the Norsemen*, Princeton, 1917.

south out of Streamfjord seeking Vinland, they still followed the shore line. After passing the southwestern extremity of Nova Scotia they did not see any land ahead in the direction in which they were going, and they could not have known that it was to the starboard afar off. Then in all probability they followed the west shore of Nova Scotia into the Bay of Fundy until they saw the island of Grand Manan or the land north of the bay. Then they probably proceeded southwest along the coast of Maine, but not very far in all likelihood, until they went ashore in what they called Hóp. There they found what they were seeking, Vinland; the proofs were there, the self-sown wheat and the grapevines.

The route that Thorfinn may have taken to Hóp has now been discussed at length. Therefore, it is not out of the way to consider some of the details regarding the sojourn there, the characteristics of the place, and the experiences encountered by Thorfinn and his men. Much has been written on that subject, and I shall mention some of it here.

Reference has been made above to the description that Jacques Cartier gave of Chaleur Bay. He affirms that wild wheat grew there in every open space in the forests; he states that the head is like rye and the kernel like oats. No doubt, he refers to the same plant as that found by Leif and Thorfinn and called wheat.

Presumably this is what is well known in America as wild rice, also as Canadian rice, Tuscarora rice, or water oats. The botanical name is *Zizania aquatica*.[49] The Indians used this as food.[50] It grows, as the Saga relates, where the land is low and wet, even in ponds; there are wide expanses of it, having the appearance of grainfields.

A number of the early European explorers of the New World mentioned grapevines that they found there. In America there are many varieties of grapevines which originally grew wild but have now been domesticated. Botanists recognize four wild varieties, *Vitis labrusca*, *V. aestivalis*, *V. cordifolia*, and *V. vulpina*. The grapes of the first and second variety are called fox grapes and doubtless are the principal kinds found by Leif and Thorfinn. The wild grapevine, although still found in some places, is now largely extinct. Babcock points out that the wild grape is now found on knolls and hillsides and until a few years ago was used quite extensively for home consumption.[51]

The Saga states that every stream was teeming with fish in Hóp, a circumstance not unnatural but worthy

---

[49] See F. C. Schübeler in *Forhandlinger i Videnskabs-selskabet*, Christiania, 1858, pp. 21-30.

[50] A. E. Jenks, The Wild Rice Gatherers of the Upper Lakes, *19th Ann. Rept. Bur. of Amer. Ethnology for 1897-98*, pp. 1013-1137.

[51] Babcock, *op. cit.*, pp. 90-94.

FIG. 20—Bluffs at Highland Light, Cape Cod. (Courtesy of American Museum of Natural History, New York.)

FIG. 21—Wild rice (*Zizania aquatica*) tied in sheaves by the Indians. (From *19th Ann. Rept. Bur. of Amer. Ethnology for 1897-98*, Pl. 72.)

of mention among the advantages of this fine new land. The method of catching halibut is rather surprising: "They dug pits on the shore where the tide rose highest, and when the tide fell there were halibut in the pits." Strange as this may seem, it is probably true. In his history of Bristol W. H. Munro gives an account of a similar method practiced there to this day. The Saga further continues: "There were many species of animals (dýr) in Hóp." The Icelandic word "dýr" has in all probability here the same meaning as "deer" in English, referring to various species of deer. Large numbers of deer formerly roamed over the east coast, and many are even found there now. Newfoundland was the home of the so-called red hart. Storm quotes an author[52] as saying that an English hunting party killed as many as two thousand of them in one summer.

The Saga gives a detailed account of the dealings that Thorfinn and his men had with the Skrælings. That narrative is too long to quote here in its entirety but may be found in Hauk's Book (pp. 439-441). Some details, however, must be noted here.

The people that Thorfinn Karlsefni and his men encountered in Hóp were named Skrælings by them, according to the Saga. Later on the name was applied to the Eskimos in Greenland. The noun "skrælingur"

[52] *Aarbøger for Nord. Oldk. og Hist.*, 1887, p. 337.

(peel, or skin) is to be found in Icelandic, as well as kindred words like "skræla" (to dry up, or peel), "skrælna" (to parch, or wither), and "skræl-thur" (parched). Ari the Learned doubtless wrote his *Islendingabók* long before the Saga of Eric the Red was penned. He says that the Greenlanders call the natives of Vinland "Skrælingar" (Skrælings) and that one can see from fragments of boats and stone implements which the Icelandic settlers, pioneers of Greenland, found there that a similar race had been there.[53] When Thorfinn was in Greenland, and even when Ari wrote his book, the Icelanders there had not encountered the Eskimos. Ari does not say that the Greenlanders called the Eskimos Skrælings, but they did so call the people that Thorfinn and his men came across in Vinland. On the other hand, when the Icelandic settlers in Greenland later met the natives and were attacked by them, they gave them the name that had been previously applied to the natives of Vinland, a name by which the Eskimos are characterized in Iceland to this day.

Now the question arises whether it is true that the same people were native to Greenland and Vinland and identical with that race which the Greenlanders met in the twelfth century and called Skrælings.

---

[53] Ari Thorgilsson, *The Book of the Icelanders*, edit. and transl. by H. Hermannsson, Ithaca, N. Y., 1930, pp. 51-52, 64.

THORFINN KARLSEFNI

There has been considerable dispute about this, but the consensus of opinion seems to be that, anthropologically speaking, this assumption is not correct. The race that was in Greenland before Eric the Red settled there was doubtless Eskimo; and to them was transferred the name "Skrælings" by the Icelanders, and they have been called so ever since. The people, however, that Thorfinn encountered in Hóp were doubtless some tribe of American Indians. Those who have written most regarding this subject are Gustav Storm[54] and W. H. Babcock,[55] and both are in agreement on that point.

The question arises whether it was incorrect to call these two by the same name, that is, consider them one and the same?

K. Birket-Smith begins his article on the Indians in "Salmonsens Konversationleksikon" thus:[56] "Indians is the common name for the aborigines of America, often excluding the Eskimos, although a sharp distinction between Indians and Eskimos is artificial."

When scientists of today are of this opinion, it does not seem inappropriate to apply the same name to the people of Vinland and Greenland. At least, the natives of those two countries were probably not much more

[54] *Aarbøger for Nord. Oldk. og Hist.*, 1887, pp. 346-355.
[55] Babcock, *op. cit.*, pp. 139-158.
[56] 2nd edit., 25 vols., Copenhagen, 1915-1927; reference in Vol. 12, p. 276.

unlike each other than are some of the Indian tribes. The name applied to them by the Icelanders and Greenlanders is much older and is not less fitting than the name Indian, which arose out of the misconception that America was India.

Thorfinn and his men came upon other aborigines on their voyage than those in Hóp, and it is not certain that the name Skrælings was first applied to the people there. This matter will be dealt with later.

At the time of Thorfinn there were doubtless many tribes of Indians, no less than now. It is not easy to determine just what tribe was in Hóp in the summer of 1004. Storm thinks that the Micmac Indians were there then. They are described in Nova Scotia six centuries later in a somewhat similar manner. Babcock believed that the Algonquins were on the New England coast in the year 1000.

The Skrælings in Hóp came from the sea in skin boats, many together. "They were evil-looking,[57] dark-skinned, with ugly hair, bulging eyes, and broad cheek bones."[58] This description does not seem alto-

[57] "Small" in Manuscript 557, 4to, is probably wrong.

[58] In the Tale of the Greenlanders the mention of this people is somewhat different, as is the case with many other details. The Tale relates that a woman, who was apparently a kind of guardian spirit of one of the Skrælings, had bulging eyes. Thus that trait given in the Tale and the Saga is doubtless a tradition found in the original story. This woman appeared only for a moment to Thorfinn's wife, Gudrid, just before the Skræling was slain. "She walked towards

gether applicable to the Indians, particularly the statement that they rowed in skin boats and had bulging eyes. Babcock has pointed out, however, that similar descriptions exist elsewhere and that some tribes of Indians used skin boats. He quotes several authorities to support his statement, both from earlier and recent times. The reference is not to kayaks, or one-man boats, but to larger ones for several men.

One of the earliest explorers of America was Giovanni da Verrazano. He is credited with a letter, dated July 8, 1524, containing a description of a voyage along the east coast. He describes one of the tribes that he saw there as having thick black hair, not very long, bound together at the back. The chief fault he finds with their features is that they are very broad-faced; not all of them, however, for some

---

Gudrid and said, 'What is your name?' 'My name is Gudrid. What is your name?' 'My name is Gudrid.'" Some writers have looked upon this as a story of a kind of doubles. One writer believed that the woman was possibly of Norse origin, although a native of Hóp. The Saga writer evidently looked upon this incident as supernatural; he describes the woman as Norse and quotes her as speaking the Norse language, thus leading one to infer that she was Norse, a strange thing, as are so many other details in the Tale. The manuscript states that this strange woman said that her name was Gudrid. That is probably a mistake in recording, a repetition of the answer that Gudrid Thorfinn's wife gave. Such mistakes, due to the carelessness of the recorder, are common. The woman's name was no doubt different. However, as this has no basis in fact, it is immaterial.

were sharp-featured and had big, dark eyes.[59] Such a description is not much at variance with the Saga. Other writers on the subject have mentioned the same facial characteristics, swarthy complexion, broad face, and bulging eyes.

The bartering that went on between Thorfinn's party and the Skrælings is in accordance with the practice of the Indians; the story relates that they came for the purpose of bartering in the spring or early in the summer. They brought furs, displayed a love of finery, and were not at all practical in their dealings. The strangers "were particularly anxious to obtain some red cloth, offered furs in exchange and totally gray pelts. They also desired to buy swords and spears, but Thorfinn and Snorri refused to sell them these. In exchange for perfect unsullied skins the Skrælings took red cloth a span in length which they would bind around their heads. This trade went on for a time, and Thorfinn and his men were getting short of cloth; then they divided it into such narrow strips that it was not more than a finger's breadth wide, but the Skrælings still continued to give them just as much as before or more."

Everything looked fairly promising for a settlement in Hóp, peaceful relations with the natives, and a productive country. However, this state of bliss

[59] See Babcock, *op. cit.*, p. 145.

THORFINN KARLSEFNI

suddenly changed. "It so happened that a bull belonging to Thorfinn ran out of the woods, bellowing loudly. The Skrælings became so terrified that they dashed to their boats and rowed southward around the shore." The peaceful bartering came to an end, and the unfortunate bellowing of Thorfinn's bull perhaps delayed the settlement of North America for five centuries.

The fact that the Skrælings were so terrified by the bull proves that they were not used to cattle. Some writers have wondered at that, since buffaloes were native to North America and large herds have been common almost up to the present time. It is not on record, however, that they were ever found on the New England coast, and Babcock is in all probability right when he states that they never penetrated the country east of the Hudson River.

The story about the bull is very probable. There are many accounts showing that timid men have been much frightened by bulls, especially when the animal was in an ugly mood. It is not likely that Thorfinn and his men did anything to allay the fear of the Skrælings; on the contrary, they possibly expressed amusement or contempt. Furthermore, their attitude in dealing with the Skrælings was probably not such as to inspire trust.

In three weeks the Skrælings came back again and

attacked Thorfinn and his men. There was a great multitude of them and a heavy shower of missiles, for the Skrælings had war slings. Thorfinn and Snorri noticed that the Skrælings raised a pole and at the end of it fastened a great ball shaped like a sheep's belly, blue in color; and this they hurled from the pole up on the land above Thorfinn's followers, and it made a terrifying noise where it fell.

It is not likely that the Skrælings knew iron or other metals, although such, including gold and copper, were known to the more civilized tribes to the south. Their war missiles were therefore natural weapons, like stones, which they slung.[60] They slew two of Thorfinn's men, Thorbrand Snorri's son[61] and another man not named. Many have thought that the North American Indians did not have war missiles. But Babcock shows that the Indians in Newfoundland had them in 1494. These he considers not unlikely to be the descendants of the Micmac tribe. Another writer was convinced that the Micmac

---

[60] These were doubtless hand slings, not the type of war slings common in the northern countries in the thirteenth century. See H. Falk, *Altnordische Waffenkunde*, Christiania, 1914, pp. 193-194.

[61] An error early crept into some manuscripts of the Eyrbyggja Saga, that Snorri himself was slain, not his son. Two of the copies of the best vellum manuscript (the Vatnshyrna), which was burnt in Copenhagen in 1728, state, however, that Thorbrand Snorri's son was slain there.

THORFINN KARLSEFNI

Indians had war missiles.[62] No doubt they were common in later centuries among the various tribes.[63]

What was the ball which the Skrælings "threw over Karlsefni and his men and which had an evil sound as it came down"? H. R. Schoolcraft, who has written an extensive work on the Indians, has explained this matter somewhat. He states that it is still remembered and quoted that the Ojibwa Indians, who were fierce warriors, used a gigantic mace in warfare, made from the pelt of an animal wrapped around a large heavy stone and put on a pole. This was brandished by a number of men, and one can imagine that it was very dangerous. According to the Saga, the ball was flung from a pole; inside of the skin there was presumably a large stone or a number of small ones. The fact that it made such an evil sound when it came down indicates that it was blown up like a bladder. It would seem that the reference was here to a very large sling rather than a mace and that its purpose was more to frighten than to wound the enemy. It is not likely that either of these weapons became generally used, rather that they soon fell into disuse among the Skrælings.

Then follows an account of the fear of Thorfinn and his men, Freydis' exhortation to them, the death

---

[62] John Fiske, *The Discovery of America*, Boston, 1893.
[63] See Babcock, *op. cit.*, pp. 154-157.

of Thorbrand Snorri's son, and how Freydis put the Skrælings to flight. The last incident is strange and incredible. "Freydis sought to follow her people, but lagged behind, because she was pregnant. She followed them, however, into the forest with the Skrælings in pursuance. She found a dead man in front of her, Thorbrand Snorri's son. His skull was split by a flat stone, and his naked sword lay beside him. She picked up the sword in order to defend herself. Then the Skrælings approached her. She uncovered her breast and slapped it with the naked sword. This action terrified the Skrælings, so they turned around and sought their boats and rowed away. Thorfinn and his men joined her and praised her good fortune." The Saga writer leads one to believe that Thorfinn and the others considered Freydis' deed rather as a piece of luck than a brave act. Her courage and strength do not need any explanation, and she must have been awe-inspiring, shaken with passionate anger as she held the naked sword in her hand. By uncovering her breast and striking it with her sword she no doubt wished to indicate two things, that she was a woman and that she was unafraid and ready to protect herself with the sharp sword if attacked. However, this does not explain the terror and flight of the Skrælings; it only explains how it happened that they did not injure her.

THORFINN KARLSEFNI

There may possibly be some misunderstanding here as to the reason for the flight of the Skrælings. Perhaps they felt that they had sufficiently avenged the ill treatment they considered themselves to have suffered; and, furthermore, they may have realized that these newcomers might become dangerous. "Two of Thorfinn's men were slain and a large number of the Skrælings." Their departure for their boats may not have been a real flight but a sudden decision to leave.

If the Skrælings were Indians, it must be remembered that these are said to be more ready to make an attack than to follow up a victory. No doubt they were superstitious and afraid of everything they did not understand or have an acquaintance with. Thus their interpretation of Freydis' way of striking her breast may have been something entirely different from anything that either she or her companions imagined. One may even infer that from the story, inasmuch as Thorfinn and his men praised her "good fortune." It is not easy to establish the real reason for the flight of the Skrælings. Doubtless Freydis made a strong impression on them and inspired them with mysterious fear; but that is hardly a sufficient reason for their flight.

It now became clear to Thorfinn and his companions that, although the country was attractive, their life would be one of dread and danger on account of unfriendly natives. This was the real reason why the scheme of colonization was abandoned.

Climate and conditions in Greenland were better then than now. There was unlimited land as in Iceland, although this had been settled for a much longer period. There had been much emigration from the mother country, Norway, during the centuries immediately preceding; therefore there was no scarcity of land there either. This situation applied also to Denmark and Sweden, to say nothing of the countries farther south, where there were great climatic advantages. For these reasons one can understand why there were no attempts to explore and settle America, although the news of the discovery of Vinland spread, as can be seen from Adam of Bremen's "Description of the Northerly Lands." He had spent some time at the court of Svein Ulfsson, king of Denmark, around 1070. The king told him about Vinland and mentioned the grapevines and the self-sown wheatfields. The Danes believed the story, as did their king; but there are no indications that it was considered of importance or given credence in other countries. No doubt the king and his court based their accounts on what some reliable Icelanders had told them, men in whom they had full confidence, probably some elderly Icelanders, who in their youth had

THORFINN KARLSEFNI

accompanied Thorfinn to Vinland or had personal acquaintance with some member of the expedition.[64]

[64] In his article in *Tímarit Thjódrœknisfélags Islendinga*, Winnipeg, Vol. 1, 1919, pp. 25-65, Halldór Hermannsson points out that the narrator on whose account the king based his statements was possibly Gellir Thorkelsson (son of Thorkel and Gudrun Osvifssdottir). According to the Laxdæla Saga, he became ill in Denmark when he was returning from Rome. He was ill for a long time and finally died, probably in Roskilde, for there he is said to be buried. Many of the Icelandic annals state that he died in 1073, and in the so-called King's Annals it is recorded that he died at the age of 65. Accordingly he was born in 1008; that is most likely, though the Laxdæla Saga intimates that he was born in 1012, saying that he was twelve years old when he first went abroad with his father. Scarcely two years later Thorkel was drowned, and the annals give that date as 1026. The Saga is not very dependable in this connection and does not agree with other reliable sources. Gellir had a son Thorgils, whose son in turn was the historian Ari the Learned. Another son of Gellir's was named Thorkel, whom Ari quotes as the one who told him about the settlement of Greenland, he having heard that story from one of the men who accompanied Eric the Red thither. No doubt that man also told Thorkel about the voyages to Vinland, whereby Gellir became well acquainted with those narratives. Adam of Bremen wrote his account, based upon the authority of King Svein, around 1070, at least before 1072. Judging from the Laxdæla Saga, Gellir had not returned to Denmark from Rome at that time. It is not likely that he came back from the south until after Adam of Bremen's visit to King Svein. Furthermore, it is not mentioned that Gellir was at the Danish Court. Another man is more likely to have given Svein and his Court the information about Vinland, namely Audun of the Westfjords, whose story is told in the vellum manuscript of the Morkinskinna and the Flatey Book. Audun had gone from the Westfjords in Iceland to Norway and thence the next summer to Greenland, where he passed a winter. The following summer he went again to Norway and from there to Denmark. He spent some time with King Svein and later went to Rome. He returned at Easter time the following year and

Although the Norwegians and Danes knew about the voyages of Thorfinn and Leif, they did not, as far as the records go, make any attempt to explore those countries or settle there. As time went on they possibly forgot all about Vinland or ceased to believe in it, since for the most part they were occupied with other matters. The Icelanders knew and recorded the accounts of Thorfinn and Leif relating to their discoveries and explorations, but none of their countrymen cared to follow their footsteps. The Icelandic colony in Greenland remembered and treasured a long time, possibly so long as it existed, these re-

---

spent some time at the Court. He is described as follows: "Audun was a man who knew how to comport himself among people with ease and grace; he was good-natured, careful of speech, and a man of few words. He was a general favorite, and King Svein was especially friendly to him." Late in the spring he went to Norway and was received by King Harald Hardredy. He went back to Iceland that summer. It is possible to ascertain the date of Audun's travels fairly accurately. According to the way in which the Tale mentions King Harald, it must have been before peace was established between him and King Svein, that is previous to 1064 or a few years before Adam of Bremen visited King Svein. It is not unlikely that King Svein asked Audun many things about Greenland, particularly since he had confidence in his reliability. It is scarcely conceivable that Audun did not tell about the Vinland voyages; in fact, he presumably quoted some of the men who were with Leif and Thorfinn. An English translation of the tale about Audun is to be found in the introduction to Sir George W. Dasent's translation of The Story of Burnt Njal (2 vols., Edinburgh, 1861), Vol. 1, pp. clxxiv-clxxxiii.

nowned voyages and endeavored, at least twice, to retrace the steps of Thorfinn and Leif. In some old Icelandic annals it is stated that Eric Gnupsson (nicknamed Pollock), bishop of the Greenlanders, went to seek Vinland in 1121. Nothing more is related about the voyage, and one may infer that Eric did not return. Three years later a new bishop was appointed in Greenland. In 1347 a small ship from Greenland was driven by storm to the coast of Iceland; there were seventeen or eighteen men on board. The ship had made a voyage to Markland. Several Icelandic annals relate that the ship went to Norway the following summer. The original voyage was probably made to get timber, but whether the ship ever reached Markland is not known.

We shall now turn again to the Saga of Eric the Red regarding Thorfinn's search for Thorhall the Hunter. It has been dealt with above (p. 25) as regards Keelness, and we considered it likely that Thorfinn went north around Cape Gaspé and west along the Gaspé Peninsula. In the second place it has been mentioned, in connection with Streamfjord, that this voyage of Thorfinn's implies that Streamfjord was not identical with the estuary of the St. Lawrence River.

It is not possible to ascertain what river is referred to in the narrative, possibly one of those far within that drain into the estuary St. Lawrence. To quote from the Saga: "When they had sailed a long time they came to a river that flowed out of the land from east to west." It is not unlikely that this river was the Ouelle River which enters the St. Lawrence from the southern part of the Kamouraska district. In the Tale of the Greenlanders it is related that Thorwald, when fatally wounded, requested his companions to bury him on a cape that was near by and was to be called Crossness. The account is not very reliable. There is, however, a cape near the Ouelle River that bears the same name as the river. The Saga describes the country as heavily wooded, with scarcely an opening, and this is in accord with the conditions of that region.

The story of the uniped that shot Thorwald, the ditty about that event, and the reference to the Land of the Unipeds are all very strange. It is possible that the ditty was originally composed during Thorfinn's voyage, and, if such is the case, the name may have originated at the same time. It is difficult to explain references to the unipeds here and elsewhere. It is not unlikely that the whole account of the uniped is a myth, an interpolation in the story. Halldór Hermannsson has drawn attention to an interesting matter in this connection, namely that Cartier had stated that Donnacona had told him in Stadaconé that he had come upon a country in which the inhabitants were unipeds.[65] The wording of the passage in the Saga implies that the country in question was northwest of the St. Lawrence.[66] Other details with regard to this voyage of Thorfinn's and the return to Streamfjord have been discussed above.

Then the Saga goes on: "The third winter that they dwelt in Streamfjord the men began to divide into factions. The women were the cause of the disagreement; those who were unmarried endeavored to seize the wives of the married men. Great trouble arose over all that. Snorri the son of Thorfinn Karlsefni was born the first autumn that they were in Streamfjord; he was three years old when the expedition took its departure."[67]

It has been mentioned above that the version of some men referred to in the Saga, to the effect that Thorfinn came back the same summer from Hóp, was more probable. But the Saga writer does not follow that here. It is possible that Thorfinn's son Snorri was three years old when they returned to Greenland, although they did not spend a winter in Hóp because they may have been two winters in Streamfjord after

[65] Hermannsson, Wineland Voyages, *Geogr. Rev.*, Vol. 17, p. 114.

[66] The words "and north again" in the Tale, referring to the flight of the uniped are doubtless an error, presumably in recording. The same words are used just afterward in regard to the departure of Thorfinn. The uniped dashed to the southern bank of the river; he had come from the south and therefore ran south again.

[67] *Hauksbók*, pp. 442-443.

the expedition to Hóp. It is not possible now to determine whether they returned to Greenland after two winters, as the Tale has it, that is in 1005, or after three winters, that is in the summer of 1006, as the Saga states.

The return voyage to Greenland is described as follows: "When they left Vinland the wind blew from the south, and they came to Markland. There they came upon five Skrælings, one bearded man, two women, and two boys. They captured the boys, but the others escaped, disappearing into the ground. They took the boys with them, and these were taught the language of their captors and baptized. They called their mother Vethilldi and their father Uvege. They said that two kings ruled the Skrælings, Avalldamon and Avalldidida. They said there were no houses in their land, the people lived in holes or caves."[68]

It would seem from this account that Thorfinn did not pursue the same route back to Markland; he may have realized that he could follow a straighter course favored by southerly winds, especially if Streamfjord was Shediac Bay. In such a case they probably went east of Anticosti Island.

[68] There is other information credited to these boys not mentioned here. Gustav Storm has written about all this at length in *Aarbøger for Nord. Oldk. og Hist.*, 1887, pp. 355ff.

The Saga does not say much about the return voyage of Thorfinn and Snorri from Markland to Greenland; but, judging from the story of Bjarni Grimulfsson and Thorhall Gamlason, who commanded the other boat and who found themselves in the "Wormy Sea," it might be inferred that the plan was to take a different course from the one followed south, that is to follow a more direct route to the Eastern Settlement. On the way back they encountered westerly winds. Thorfinn and his men reached Greenland, but the other ship became so damaged that it sank with probably most of those on board, inasmuch as only a few could find room in the lifeboat. Thorhall Gamlason was among those who were rescued, and after that he was always called "the Vinlander." He lived afterwards at Melar in Hrutafjord, Iceland; and his sons are connected with the Saga of Grettir the Strong.

Many have written at length about the nationality of the little family that Thorfinn and his companions met in Markland. One of the outstanding treatises on the subject is by William Thalbitzer of the University of Copenhagen.[69] He declares that the alleged proper names of the natives of Markland are not names of persons, but that the boys had referred to

[69] William Thalbitzer, Skrælingerne i Markland og Grønland, *Oversigt over det Kgl. Danske Vidensk. Selsk. Forhandl.*, 1905, No. 2, pp. 158-209.

something else in these or similar words. The first word means "stop now a moment"; the second, "stop a moment now"; the third, "out to the outermost (land)"; the fourth, "you mean the outermost?" It is not likely that the Icelanders and these boys from Markland understood each other very well. It is probable also that the words ascribed to the boys were not repeated and preserved on record without some change.

Thalbitzer brings out many arguments to support the contention that the name of the natives, "Skrælings," had its origin among the people themselves, as a number of others also have believed. A group of Eskimos in Labrador (Helluland) called another tribe north of them "Karaleq," about 1760; and by that name the Skrælings who were in the old settlements in Greenland called themselves. One must remember that the Icelanders could not have pronounced the name "Karaleq" the same way as the Eskimos,[70] nor would it have been spelt phonetically correct in Icelandic. Very likely it was connected with similarly sounding words in Icelandic which have been mentioned above.

The narrative about Thorfinn's voyage to Vinland has now come to an end, so far as it is related in the

[70] The word Eskimos came into French, borrowed from the Indians north of the Gulf of St. Lawrence (Wonderstrands); they called thus those who lived north of them. It signifies "eaters of raw meat."

Saga. The beginning of the last chapter is as follows: "The second summer (after landing in Greenland) Thorfinn and his wife Gudrid went to Iceland." The Tale of the Greenlanders states that they went from Greenland to Norway in the summer after their return from Vinland, spent the winter there, sold their goods, and proceeded the next summer to Iceland. It is probable that the second version is nearer the truth, for it is likely that Thorfinn had gathered quite a cargo on his voyage to Greenland and Vinland, goods that he could dispose of more easily in Norway than in Iceland.

The story of Thorfinn in Hauk's Book begins with his genealogy and ends with a list of his descendants. He was a good representative of a great family. His father, Thord Horsehead, was a grandson of one of the settlers of Iceland, who bore the same name and who was the fourth or fifth man in direct male line from the famous Ragnar Lodbrok, and whose wife was the granddaughter of Kjarval, king of Ireland. Thorfinn's grandmother on his father's side was the daughter of Thord Gellir, who was a direct descendant of kings, one after the other. Among Thorfinn's descendants were Bishop Thorlak Runolfsson and Bishop Björn Gislason, both his great-grandsons, while Bishop Brand Sæmundsson was his great-great-grandson.

It proved to be too great a task for Thorfinn to establish a colony in the New World. He himself was neither a Viking nor a warrior king. He was not in a position to seek aid from either earls or kings for his project. The Faeroes, Iceland, and Greenland were unsettled and undefended countries when Grim Kamban, Ingolf Arnarson, and Eric the Red established themselves there. In contrast, Vinland was inhabited by a multitude of armed tribes, unquestionably accustomed to internal warfare. The nations that stood behind Thorfinn were small and, relatively speaking, poor. They had more land at their disposal than they could handle. The new land could be reached only by a long voyage over a wide expanse of open sea, fraught with hardships and difficulties. The Danes and the Swedes had established colonies in countries that were difficult to settle; but they did not have to cross a vast ocean, and great wealth and varied products were to be had near at hand. In Vinland, on the other hand, everything had to be taken out of a virgin soil. The natives were savage nomads, who lived by fishing and hunting. Many centuries later the Europeans had to fight hard before they could form permanent settlements in the New World. It was not until the last century that some of the descendants of Thorfinn and his companions established colonies in that hemisphere to which Vinland the Good belongs,

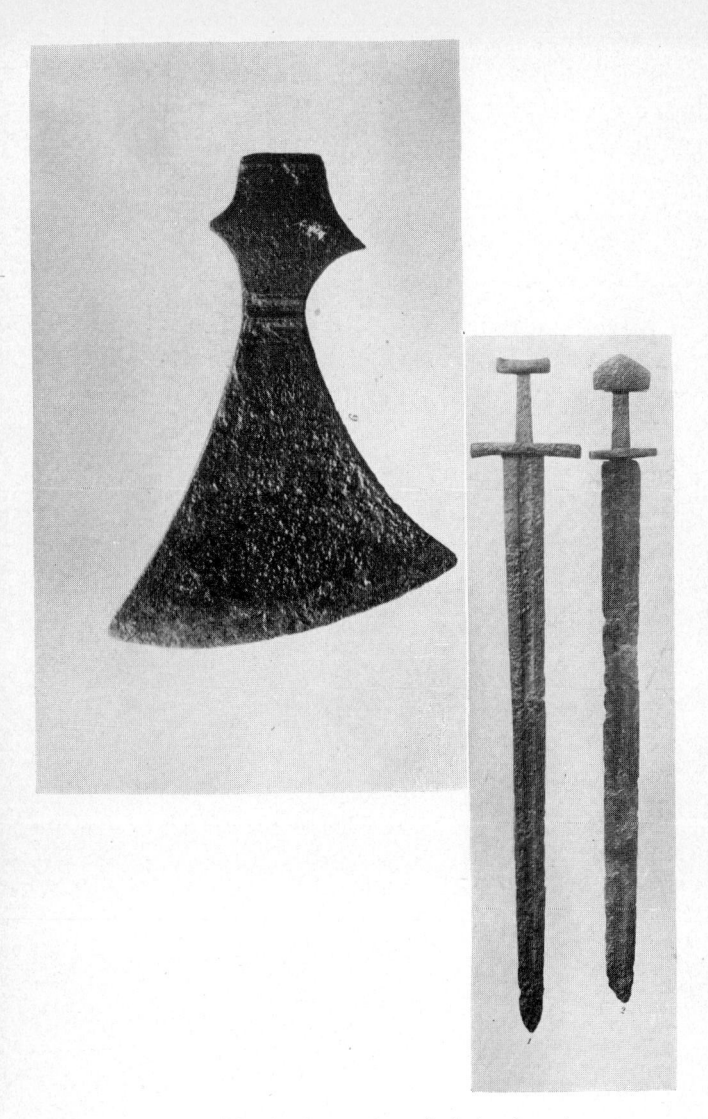

FIGS. 22 and 23—Viking battle-ax and swords dating from the year 1000. (National Museum of Iceland, Reykjavik.)

CONCLUSION

colonies that are larger in extent and population than would in all likelihood have been possible in Thorfinn's time.

Nevertheless, Thorfinn's brave attempt lives in the memory of many of those who dwell in Vinland today. In recent times several noted men and women of the New World have made serious researches as to his voyage, written on the subject, and studied all the Icelandic source material carefully. In 1920 a dignified bronze monument was erected to his memory in Philadelphia, through the generosity of J. Bunford Samuel (see the frontispiece). The sculptor was the renowned Icelandic artist Einar Jónsson. No one, however, has ever advanced the theory that Thorfinn reached Philadelphia. The voyage of Leif Ericsson had already been commemorated in a similar manner many years before, by the erection in 1886 in the

ADDITIONAL FOOTNOTE TO P. 38.—Fernald identifies the grapevine (*vinber*) and self-sown wheat of the Sagas as the mountain cranberry (*Vaccinium Vitis-Idæa*) and strand wheat (*Elymus arenarius*) respectively. The area of their greatest abundance, he states, lies north of the lower St. Lawrence along the coast of Labrador. His conclusion is: "If, however, the newer interpretation is admitted, that mountain cranberries, birch trees, and strand wheat were what the Norsemen really saw in Markland and Vinland, it will be noted that in Newfoundland and Labrador every one of the natural features is satisfied" (*Bull. Amer. Geogr. Soc.*, Vol. 47, 1915, p. 687). But his view of the matter has not been generally accepted, except in so far as the absence of wild grapes in Nova Scotia is concerned. Hence the attempts at seeking Vinland farther south.—H. H.

Fenway, Boston, of a statue of him by Anne Whitney.

The location of Hóp has not yet been ascertained, and the position of Streamfjord and other places mentioned in the narrative of Thorfinn's voyage is still a matter of doubt. The research work should continue and, as Halldór Hermannsson and Finnur Jónsson have emphasized, be carried on by investigating those places which Leif and Thorfinn most likely visited on their voyages.

INDEX

INDEX

Adam of Bremen, 56, 57
Algonquins, 48
Alptafjord, 11
Althing, 2
America, xii; discovery, xiv; settlement abandoned, 56
Animals, wild, 34, 36, 45
Anticosti Island, 62
Arctic foxes, 13, 18
Ari Thorgilsson, 4, 46, 57
Arna-Magnaean Collection, 2; Codex No. 544 (facsimiles), opp. 10, opp. 34
Audun of the Westfjords, 57
Avalldamon, 62
Avalldidida, 62

Babcock, W. H., 40, 41, 42, 44, 47, 48, 49, 50, 51, 52, 53
Ball and pole, 52, 53
Barter, 35, 45, 50
Battle-ax of Vikings (ill.), opp. 66
Bay of Islands, Newfoundland (ill.), opp. 22
Bays, 14, 23, 26
Bear Island, 13, 14; location, 15, 16
Belle Isle, 19, 21
Belle Isle, Strait of, 8, 19, 22
Berries, 29
Birds, 28
Birket-Smith, Kaj, 47
Bjarni Grimulfsson, 11, 12, 13, 34, 36, 63
Björn Gislason, 65
Blanc Sablon, 19
Bon Ami, Cape (ill.), opp. 24
Bonaventure Island (ill.), opp. 30; bird ledges on (ill.), opp. 28
Boston, 40, 68
Brand Jónsson, 3

Brand Sæmundsson, 3, 65
Brattahlid, 6, 9, 11
Breidafjord, 11
Bristol, R. I., 42, 45
Bristol Narrows, 41
Britain, vii
Buffaloes, 51
Bull, 35, 51

Cabot Strait, 9
Cambridge, Mass., 40
Canadian rice, 44
Canoes, skin, 34, 35, 49
Canso, Strait of, 37
Cape Breton, 37
Cape Cod, 42; bluffs at Highland Light (ill.), opp. 44
Caraquet Bay, 30
Cartier, Jacques, 29, 33, 43, 60
Cattle, 51
Chaleur Bay, 28, 29, 30, 36, 37, 43; head (ill.), opp. 28
Charles River, 40
Christianity, 1, 2
Christians, vi
Climate, 56
Cloth, 50
Coast line, sinking of, 40, 41
Cocagne Harbor, 37
Colonization, 15, 56, 66
Coracles, viii
Crossness, 20 (map), 60

Dægr, 13, 16
Danes, 56, 58, 66
Dasent, Sir G. W., 58
Deer, 45
Denmark, 56, 57
Dicuil, ix
Directions, 21

Disko, 15
Ditties, 32, 33, 60
Donnacona, 60
Ducks, eider, 14, 28
Dyr, 45

Eastern Channel, 41
Eider ducks, 14, 28
Elymus arenarius, 67
England, xi
Ere-Dwellers, Saga of the, 11
Eric Gnupsson, 59
Eric the Red, xiv, 1, 9, 11, 13, 66; home in Iceland (ill.), opp. 2; Saga of, 2, 3, 5
Ericsfjord, 6, 9; outlook from Eric the Red's home (ill.), opp. 6
Erlingsson, Thorsteinn, 40
Eskimos, 45; Indians and, 47; name, 64
Eyrbyggja Saga, 11, 52

Faeroes, ix, xi, 6, 66
Falk, H., 52
Fernald, M. L., 38, 67
Fish, 14, 15, 29, 34, 44; catching, 45
Fiske, John, 53
Flat stones, 13, 18
Flatey Book, 5, 57
Flateyjarbók, 5
Flatstoneland, 13, 18
Forestland, 14
Fox grapes, 44
France, vi, x
Freydis, 13, 35, 53, 55
Fundy, Bay of, 43
Furdustrandir, 14, 21
Furs, 50

Gaspé, Bay of, 26
Gaspé, Cape, 25, 30, 36, 59; extreme point and secondary headland, Cape Bon Ami (ills.), opp. 24

Gaspé Peninsula, 25, 26, 28, 37, 59; bird ledges, Bonaventure Island (ill.), opp. 28; cape at Percé showing Percé Rock and Bonaventure Island (ill.), opp. 30
Gathorne-Hardy, G. M., 27, 39
Gellir Thorkelsson, 57
Gissur Teitsson, 1, 2
Grain, 29
Grand Manan, 43
Grapes, 31, 33, 34; kinds, 44
Grapevines, 6, 7, 9, 29, 33, 38, 39, 43, 56, 67; kinds, 44
Green Island, 28
Greenland, v, 45, 46, 56, 66; Eastern Settlement, 6, 9, 15, 63; Icelanders in, 58; name, xv; return voyage to, 62, 63; Western Settlement, 11, 13, 15
Grettir the Strong, Saga of, 63
Grim Kamban, 66
Gudmundsson, Valtyr, 40
Gudrid, 4, 10, 11, 15, 36, 48, 65
Gudrid (supernatural being), 49
Gunnbjörn, xiv

Hækja, 14, 26
Haki, 14, 26
Halibut, 34, 45
Hall Gissursson, 4
Harald Hardredy, 58
Harald the Fairhaired, x
Hare Island, 27
Hauk Erlendsson, 2
Hauk's Book, 2, 5, 9, 65; facsimile of page containing description of Hóp (ill.), opp. 34; facsimile of story of Leif Ericsson's finding of Vinland (ill.), opp. 10; narratives analyzed, 12
Hauksbók, 2
Hebrides, 6
Helgafell, 4
Helluland, 13, 16, 18, 42, 64
Hermannsson, Halldór, 19, 21, 25, 33, 57, 68; on Streamfjord, 28, 30,

INDEX

36; on unipeds, 60; on Vinland and Hóp, 39
Heron Island, 30
Highland Light, Cape Cod (ill.), opp. 44
Hjalti Skeggjason, 1, 2
Hólar, Bishop of, 3
Holm, Gustav, 27, 36; on Vinland and Hóp, 39
Hóp, 24, 25, 27, 28, 33, 34, 36, 37, 61; description in Hauk's Book (facsimile), opp. 34; Indians in, 48, 51; location, 38, 41, 43, 68; name, 42
Horn, The, Iceland (ill.), opp. 2
Horsford, Cornelia, 40, 41
Horsford, E. N., 40, 41
Hrutafjord, 63
Hudson River, 39, 40, 51

Iceland, v, 56, 65, 66; discovery, ix, xi; Irish in, xii; population, xii, xiii
Icelanders, 56; in Greenland, 46, 58
Icelandic Book of Settlement, 12
Indians, 29, 42, 44, 47, 50; name, 47, 48; Skrælings as, 55; weapons, 52
Ingolf Arnarson, 66
Ireland, viii, xi, 32
Irish, viii, xi; discovery of Iceland, ix; in Iceland, xii
Isle de Bacchus, 33
Isle d'Orléans, 33
Islendingabók, 46

Jenks, A. E., 44
Jón Thórdarson, 5
Jónsson, Arngrim, 7
Jónsson, Einar, 67; statue of Karlsefni (ill.), frontispiece
Jónsson, Eiríkur, 3
Jónsson, Finnur, 3, 5, 16, 26, 68

Kamouraska district, 60
Karaleq, 64

Karlsefni, Thorfinn, 4, 11, 12, 13, 34, 37, 58, 65; conjectural route of voyage, 1003-1006 (map), 17; detail of conjectural route in the Gulf of St. Lawrence (map), 20; expedition of 1004, 31; family, 65; monument to, 67; return voyage to Greenland, 62, 63; search for Thorhall, 59; statue in Philadelphia, frontispiece (ill.), 67; voyage as influenced by Leif's, 22
Kayaks, 49
Keelness, 14, 19, 30, 32, 37; location, 22, 25, 36
Ketil Hermundsson, 4
Kjarval, 65
Krossanes. See Crossness

Labrador, 18, 21, 27; coast near Cape Mugford (ills.), opp. 18
Landnámabók, 12
Laugabrekka, 10
Laxdæla Saga, 57
Leif Ericsson, 1, 2, 10, 58; conjectural route of voyage in the year 1000 (map), 8; discovery of a new world, 2; influence of his voyage on Thorfinn's, 22; probable birthplace (ill.), opp. 2; statue in Boston, 67-68; voyage from Norway and discoveries in 1000, 5, 6; voyage to Norway, 6, 8 (map)
Livestock, 14, 34, 35
Long Island, 40
Long Island Sound, 39
Lysufjord, opp. 10 (ill.), 11, 15, 16

Mace, 53
Magnússon, Arni, 2
Maine, 39, 42, 43
Malbaie, 26
Markland, 14, 16, 59; natives, 62, 63; Newfoundland as, 18; Skrælings in, 62; Thorfinn's voyage to, 18

Maryland, 7
Mausur, 6, 8
Melar, 63
Metals, 52
Micmac Indians, 48, 52-53
Migration, x
Miramichi Bay, 37
Miscou Island, 28, 30
Montmagny, 25, 27
Morkinskinna, 57
Mount Hope Bay, 41
Mountain cranberry, 67
Mugford, Cape, coast near (ills.), opp. 18
Munro, W. H., 42, 45

Naddodd, xi
Narragansett Bay, 41
New Brunswick, 7, 25, 29, 30, 37, 39
New England, 7, 38, 39
New World, difficulty of settlement, 66; discovery, xii, xiv
New York, 40
New York Bay, 39, 40
Newfoundland, 7, 9, 22, 27, 45, 52; as Markland, 18; Bear Island and, 19; coast (ills.), opp. 22
Nidaros (Trondhjem), 6; harbor headland (ill.), opp. 2
Normandy, vi, x
Norsemen, v, vi, x
North America, 7
Norway, viii, x, 56, 65
Norwegians, viii, 58
Notre Dame Bay, Newfoundland (ill.), opp. 22
Nova Scotia, 7, 9, 22, 36, 37, 38, 40, 43, 48, 67

Ojibwa Indians, 53
Olaf Tryggvason, 1; Saga of, 5
Ouelle River, 60

Passamaquoddy Bay, 39
Percé Rock (ill.), opp. 30
Philadelphia, 67; statue of Thorfinn Karlsefni (ill.), frontispiece
Pilgrim Islands, 28
Pits for catching fish, 34, 45
Point des Monts, 23, 26, 28
Point Natashkwan, 24
Point Vaches, 23, 25
Pole and ball, 52, 53
Polygamy, xiii
Pottery, 41
Prince Edward Island, 37
Pytheas, viii

Quebec, 27

Rafn, C. C., 38, 42
Ragnar Lodbrok, 65
Red hart, 45
Reeves, A. M., 3, 5, 31
Rhode Island, 41
Rice, wild, 44
Rivière du Sud, 27
Rome, 57
Roses, 29
Roskilde, 57
Ruins, Norse, 40, 41; in Greenland (ill.), opp. 6

Sagas, 2
Saguenay River, 23
St. Lawrence, Gulf of, 20 (map), 21
St. Lawrence River, 23, 27, 31, 33, 40; estuary, 27, 37, 59, 60
St. Thomas, 27
Sakonnet River, 41
Salmon, 29
Samuel, J. B., 67
Schoolcraft, H. R., 53
Schübeler, F. C., 44
Scotland, viii
Scots, viii
Seven Islands, 21
Shediac Bay, 37, 62
Shetlands, 6
Shippigan Island, 28
Ship's keel, 14, 23
Sinking of coast, 40, 41

INDEX

Skagafjord, 11
Skálholt Cathedral, 3
Skerries, xiv
Skin boats, 34, 35, 49
Skrælingar, 46
Skrælings, 34, 35; dealings with, 35, 45, 50; description, 48; fright and flight, 35, 51, 54, 55; identification, 46; in Hóp, 48; Markland, 62; name, 45, 47, 48, 64
Slavery, xiii, 33
Slings, 52, 53
Snæfellsness, 4, 10
Snorri, son of Thorfinn Karlsefni, 61
Snorri Thorbrandsson, 11, 12, 13, 34, 52
Stadaconé, 60
Steensby, H. P., 15, 27, 33; on Bear Island, 19; on Markland, 18; on Vinland and Hóp, 39, 42; on Wonderstrands, 21, 27
Stefánsson, Vilhjálmur, v
Storm, Gustav, 3-4, 5, 6, 16, 45, 47, 62; on Newfoundland as Markland, 18; on Vinland, 38; on Wonderstrands, 21
Strand wheat, 67
Strands, 14
Straumey, 14
Straumfjord, 14
Stream Island, 14, 27
Streamfjord, 14, 24, 25, 27, 36, 59; location, 26, 28, 30, 37, 68; troubles in, 61
Sturla Thórdarson, 4
Summer day in the North, ix
Svein Ulfsson, 56, 57
Svold, Island of, 2
Sweden, 56
Swedes, 66
Swords of Vikings (ill.), opp. 66

"Tale of Eric the Red," 5
"Tale of the Greenlanders," 4, 5, 18, 48
Taunton River, 41
Thalbitzer, William, 63, 64

Thingvellir, xiii
Thorbjörn Vifilsson, 10, 11
Thorbrand, 52, 54
Thord Gellir, 65
Thord Horsehead, 65
Thorfinn Thórdarson, 11. *See also* Karlsefni
Thorgils, 57
Thorhall Gamlason, 11, 12, 13, 63
Thorhall the Hunter, 13, 23, 24, 28, 30, 36; death, 32; ditties, 32, 33; expedition of 1004, 31; Thorfinn's search for, 30, 59
Thorkel, 57
Thorlak Runolfsson, 65
Thorstein Ericsson, 9, 10
Thorvald Asvaldsson, xiv
Thorwald Ericsson, 24, 60
Thorward, 13
Torfason, Thermód, 7
Trondhjem. *See* Nidaros
Tuscarora rice, 44

Ulfljot, xiii
Unger, C. R., 5
Unipeds, 24, 60, 61; Land of the, 24, 60
Uvege, 62

Vaccinium Vitis-Idæa, 67
Verrazano, Giovanni da, 49
Vethilldi, 62
Vídalín, Jón, 7
Vigfusson, G., 5
Viking Age, v
Vikings, vi; battle-ax and swords (ill.), opp. 66; discoveries, xi; seamanship, vii
Vinber, 67
Vinland, 1, 9, 56, 57, 58, 59, 62, 66; expedition to seek, in 1004, 31; name, 12
Vinland the Good, 12
Vinland voyages, 4
Vinlander, 63
Vitis sp., 44

War slings, 52, 53
Water oats, 44
Weapons, 50, 52, 53; Viking (ill.), opp. 66
Wends, Land of the, 1
Westfjords, 57
Whale, 14
Wheat, self-sown, 6, 9, 34, 38, 39, 43, 56
White sands, 18, 19

Whitney, Anne, 68
Whittle, Cape, 21, 23
Wild grape, 44, 67
Wild rice, 44, opp. 44 (ill.)
Wine, 32, 33
Wonderstrands, 14, 21, 23, 26, 31, 32; as Labrador, 27
Woods, 29
Wormy Sea, 63

Zizania aquatica, 44, opp. 44 (ill.)

Augsburg College
George Sverdrup Library
Minneapolis, Minnesota 55404